International Acclaim for Martin Rees's
OUR FINAL HOUR

OUR FINAL HOUR

A Scientist's Warning:
How Terror, Error, and
Environmental Disaster
Threaten Humankind's Future
in This Century—
On Earth and Beyond

Martin Rees

BASIC
BOOKS

A Member of the Perseus Books Group
New York

Published by Basic Books,
A Member of the Perseus Books Group

Hardback first published in 2003 by Basic Books
Paperback first published in 2004 by Basic Books

Books published by Basic Books are available at special discounts for bulk
purchases in the United States by corporations, institutions, and other
organizations. For more information, please contact the Special Markets
Department at the Perseus Books Group, 11 Cambridge Center,
Cambridge, MA 02142, or call (617) 252-5298, (800) 255-1514 or e-mail
special.markets@perseusbooks.com.

Library of Congress Cataloging-in-Publication Data
Rees, Martin J., 1942–
Our final hour: A scientist's warning: How terror, error, and
environmental disaster threaten humankind's future in this century—
on Earth and beyond / Martin Rees.
p. cm.
Includes bibliographical references and index.
ISBN 0-465-0682-6 (hc); ISBN 0-465-06863-4 (pbk)
1. Twenty-first century—Forecasts. 2. Disasters—Forecasts.
3. End of the world. 1. Title.
CB161.R38 2003
303.49'09'05—dc21
2003000301

Design by Jane Raese
Text set in 11 point Janson

05 06 / 10 9 8 7 6 5 4 3

CONTENTS

PREFACE

SCIENCE IS ADVANCING FASTER THAN EVER, and on a broader front: bio-, cyber- and nanotechnology all offer exhilarating prospects; so does the exploration of space. But there is a dark side: new science can have unintended consequences; it empowers individuals to perpetrate acts of megaterror; even innocent errors could be catastrophic. The "downside" from twenty-first century technology could be graver and more intractable than the threat of nuclear devastation that we have faced for decades. And human-induced pressures on the global environment may engender higher risks than the age-old hazards of earthquakes, eruptions, and asteroid impacts.

This book, though short, ranges widely. Separate chapters can be read almost independently: they deal with the arms race, novel technologies, environmental crises, the scope and limits of scientific invention, and prospects for life beyond the Earth. I've benefited from discussions with many specialists; some of them will, however, find my cursory presentation differently slanted from their personal assessment. But these are controversial themes, as indeed are all "scenarios" for the long-term future.

If nothing else, I hope to stimulate discussion on how to

guard (as far as is feasible) against the worst risks, while deploying new knowledge optimally for human benefit. Scientists and technologists have special obligations. But this perspective should strengthen everyone's concern, in our interlinked world, to focus public policies on communities who feel aggrieved or are most vulnerable.

I thank John Brockman for encouraging me to write the book. I'm grateful to him and to Elizabeth Maguire for being so patient, and to Christine Marra and her colleagues for their efficient and expeditious efforts to get it into print.

OUR FINAL HOUR

I

PROLOGUE

THE TWENTIETH CENTURY BROUGHT US THE BOMB, and the nuclear threat will never leave us; the short-term threat from terrorism is high on the public and political agenda; inequalities in wealth and welfare get ever wider. My primary aim is not to add to the burgeoning literature on these challenging themes, but to focus on twenty-first century hazards, currently less familiar, that could threaten humanity and the global environment still more.

Some of these new threats are already upon us; others are still conjectural. Populations could be wiped out by lethal "engineered" airborne viruses; human character may be changed by new techniques far more targeted and effective than the nostrums and drugs familiar today; we may even one day be threatened by rogue nanomachines that replicate catastrophically, or by superintelligent computers.

Other novel risks cannot be completely excluded. Experiments that crash atoms together with immense force could start a chain reaction that erodes everything on Earth; the experiments could even tear the fabric of space itself, an ultimate "Doomsday" catastrophe whose fallout spreads at the speed of light to engulf the entire universe. These latter scenarios may be exceedingly unlikely, but they raise in extreme form the issue of who should decide, and how, whether to proceed with experiments that have a genuine scientific purpose (and could conceivably offer practical benefits), but that pose a very tiny risk of an utterly calamitous outcome.

We still live, as all our ancestors have done, under the threat of disasters that could cause worldwide devastation: volcanic supereruptions and major asteroid impacts, for instance. Natural catastrophes on this global scale are fortunately so infrequent, and therefore so unlikely to occur within our lifetime, that they do not preoccupy our thoughts, nor give most of us sleepless nights. But such catastrophes are now augmented by other environmental risks that we are bringing upon ourselves, risks that cannot be dismissed as so improbable.

During the Cold War years, the main threat looming over us was an all-out thermonuclear exchange, triggered by an escalating superpower confrontation. That threat was apparently averted. But many experts—indeed, some who themselves controlled policy during those years—believed that we were lucky; some thought that the cumulative risk of Armageddon over that period was as much as fifty percent. The immediate danger of all-out nuclear war has receded. But there is a growing threat of nuclear weapons being used sooner or later somewhere in the world.

Nuclear weapons can be dismantled, but they cannot be uninvented. The threat is ineradicable, and could be resurgent in the twenty-first century: we cannot rule out a realignment that would lead to standoffs as dangerous as the Cold War rivalry, deploying even bigger arsenals. And even a threat that seems, year by year, a modest one mounts up if it persists for decades. But the nuclear threat will be overshadowed by others that could be as destructive, and far less controllable. These may come not primarily from national governments, not even from "rogue states," but from individuals or small groups with access to ever more advanced technology. There are alarmingly many ways in which individuals will be able to trigger catastrophe.

The strategists of the nuclear age formulated a doctrine of deterrence by "mutually assured destruction" (with the singularly appropriate acronym MAD). To clarify this concept, real-life Dr. Strangeloves envisaged a hypothetical "Doomsday machine," an ultimate deterrent too terrible to be unleashed by any political leader who was one hundred percent rational. Later in this century, scientists might be able to create a real nonnuclear Doomsday machine. Conceivably, ordinary citizens could command the destructive capacity that in the twentieth century was the frightening prerogative of the handful of individuals who held the reins of power in states with nuclear weapons. If there were millions of independent fingers on the button of a Doomsday machine, then one person's act of irrationality, or even one person's error, could do us all in.

Such an extreme situation is perhaps so unstable that it could never be reached, just as a very tall house of cards, though feasible in theory, could never be built. Long before individuals acquire a "Doomsday" potential—indeed, perhaps within a decade—some will acquire the power to trigger, at unpredictable times, events on the scale of the worst present-day terrorist outrages. An organised network of Al Qaeda-type terrorists would not be required: just a fanatic or social misfit with the mindset of those who now design computer viruses. There are people with such propensities in every country—very few, to be sure, but bio- and cyber-technologies will become so powerful that even one could well be too many.

By mid-century, societies and nations may have drastically realigned; people may live very differently, survive to a far greater age, and have different attitudes from those of the present (maybe modified by medication, chip implants, and so forth). But one thing is unlikely to change: individuals will make mistakes, and there will be a risk of malign actions by embittered loners and dissident groups. Advanced technology will offer new instruments for creating terror and devastation; instant universal communications will amplify their societal impact. Catastrophes could arise, even more worryingly, simply from technical misadventure. Disastrous accidents (for instance, the unintended creation or release of a noxious fast-spreading pathogen, or a devastating software error) are possible even in well-regulated institutions. As the threats become graver, and the possible perpetrators more numerous, disruption may become so pervasive that society corrodes and regresses. There is a longer-term risk even to humanity itself.

Science is emphatically not, as some have claimed, approaching its end; it is surging ahead at an accelerating rate. We are still flummoxed about the bedrock nature of physical reality, and the complexities of life, the brain, and the cosmos. New discoveries, illuminating all these mysteries, will engender benign applications; but will also pose new ethical dilemmas and bring new hazards. How will we balance the multifarious prospective benefits from genetics, robotics, or nanotechnology against the risk (albeit smaller) of triggering utter disaster?

My special scientific interest is cosmology: researching our environment in the widest conceivable perspective. This might seem an incongruous viewpoint from which to focus on practical terrestrial issues: in the words of Gregory Benford, a fiction writer who is also an astrophysicist, study of the "grand gyre of worlds. . . imbues, and perhaps afflicts, astronomers with a per-

ception of how like mayflies we are." But few scientists are un-
worldly enough to fit Benford's description: a preoccupation
with near-infinite spaces doesn't make cosmologists especially
"philosophical" in coping with everyday life; nor are they less
engaged with the issues confronting us here on the ground, today and tomorrow. My subjective attitude was better expressed
by the mathematician and philosopher Frank Ramsey, a member of the same College in Cambridge (King's) to which I now
belong: "I don't feel the least humble before the vastness of the
heavens. The stars may be large, but they cannot think or love;
and these are qualities which impress me far more than size
does. . . . My picture of the world is drawn in perspective, and
not like a model drawn to scale. The foreground is occupied by
human beings, and the stars are all as small as threepenny bits."

A cosmic perspective actually strengthens our concerns
about what happens here and now, because it offers a vision of
just how prodigious life's future potential could be. Earth's
biosphere is the outcome of more than four billion years of
Darwinian selection: the stupendous time spans of the evolutionary past are now part of common culture. But life's future
could be more prolonged than its past. In the aeons that lie
ahead, even more marvellous diversity could emerge, on and
beyond Earth. The unfolding of intelligence and complexity
could still be near its cosmic beginnings.

A memorable early photograph taken from space depicted
"Earthrise" as viewed from a spacecraft orbiting the Moon.
Our habitat of land, oceans, and clouds was revealed as a thin
delicate glaze, its beauty and vulnerability contrasting with the
stark and sterile moonscape on which the astronauts left their
footprints. We have had these distant images of the entire
Earth only for the last four decades. But our planet has existed
for more than a hundred million times longer than this. What
transformations did it undergo during this cosmic time span?

About 4.5 billion years ago our Sun condensed from a cosmic cloud; it was then encircled by a swirling disk of gas. Dust in this disk agglomerated into a swarm of orbiting rocks, which then coalesced to form the planets. One of these became our Earth: the "third rock from the Sun." The young Earth was buffeted by collisions with other bodies, some almost as large as the planets themselves: one such impact gouged out enough molten rock to make the Moon. Conditions quietened and Earth cooled. The next transformations distinctive enough to be seen by a faraway observer would have been very gradual. Over a prolonged time span, more than a billion years, oxygen accumulated in Earth's atmosphere, a consequence of the first unicellular life. Thereafter, there were slow changes in the biosphere, and in the shape of the land masses as the continents drifted. The ice cover waxed and waned: there might even have been episodes when the entire Earth froze over, appearing white rather than pale blue.

The only abrupt worldwide changes were triggered by major asteroid impacts or volcanic supereruptions. Occasional incidents like these would have flung so much debris into the stratosphere that for several years, until all the dust and aerosols settled again, Earth looked dark grey, rather than bluish white, and no sunlight penetrated down to land or ocean. Apart from these brief traumas, nothing happened suddenly: successions of new species emerged, evolved, and became extinct on geological time scales of millions of years.

But in just a tiny sliver of Earth's history—the last one-millionth part, a few thousand years—the patterns of vegetation altered much faster than before. This signalled the start of agriculture: the imprint on the terrain of a population of humans, empowered by tools. The pace of change accelerated as human populations rose. But then quite different transformations were perceptible, and these were even more abrupt.

Within fifty years, little more than one hundredth of a millionth of Earth's age, the amount of carbon dioxide in the atmosphere, which over most of Earth's history had been slowly falling, began to rise anomalously fast. The planet became an intense emitter of radio waves (the total output from all TV, cellphone, and radar transmissions).

And something else happened, unprecedented in Earth's 4.5 billion year history: metallic objects—albeit very small ones, a few tonnes at most—left the planet's surface and escaped the biosphere completely. Some were propelled into orbits around Earth; some journeyed to the Moon and planets; a few even followed a trajectory that would take them deep into interstellar space, leaving the solar system for ever.

A race of scientifically advanced extraterrestrials watching our solar system could confidently predict that Earth would face doom in another six billion years, when the Sun, in its death throes, swells up into a "red giant" and vaporises everything remaining on our planet's surface. But could they have predicted this unprecedented spasm less than halfway through Earth's life—these human-induced alterations occupying, overall, less than a millionth of our planet's elapsed lifetime and seemingly occurring with runaway speed?

If they continued to keep watch, what might these hypothetical aliens witness in the next hundred years? Will a final squeal be followed by silence? Or will the planet itself stabilise? And will some of the small metallic objects launched from Earth spawn new oases of life elsewhere in the solar system, eventually extending their influences, via exotic life, machines, or sophisticated signals, far beyond the solar system, creating an expanding "green sphere" that eventually pervades the entire Galaxy?

It may not be absurd hyperbole—indeed, it may not even be an overstatement—to assert that the most crucial location in space and time (apart from the big bang itself) could be here and now. I think the odds are no better than fifty–fifty that our present civilisation on Earth will survive to the end of the present century. Our choices and actions could ensure the perpetual future of life (not just on Earth, but perhaps far beyond it, too). Or in contrast, through malign intent, or through misadventure, twenty-first century technology could jeopardise life's potential, foreclosing its human and posthuman future. What happens here on Earth, in this century, could conceivably make the difference between a near eternity filled with ever more complex and subtle forms of life and one filled with nothing but base matter.

2

TECHNOLOGY SHOCK

Twenty-first century science may alter human beings themselves—not just how they live. A superintelligent machine could be the last invention humans ever make.

"IN THE PAST CENTURY, there were more changes than in the previous thousand years. The new century will see changes that will dwarf those of the last" This was an oft-expressed sentiment in the years 2000 and 2001, at the dawn of the new millennium; but these words actually date from more than one hundred years ago, and refer to the nineteenth and twentieth centuries, not the twentieth and twenty-first. They are from a 1902 lecture entitled "Discovery of the Future" presented by the young H.G. Wells at the Royal Institution in London.

By the end of the nineteenth century, Darwin and the geologists had already delineated, in crude outline, how Earth and its biosphere had evolved. Earth's full age was still not recognised, but estimates had risen to hundreds of millions of years. Wells himself was taught these ideas, still novel and inflammatory at that time, by Darwin's greatest advocate and propagandist, T.H. Huxley.

Wells's lecture was mainly in visionary mode. "Humanity," he said, "has come some way, and the distance we have travelled gives us some earnest of the way we have to go. All the past is but the beginning of a beginning; all that the human mind has accomplished is but the dream before the awakening." His rather purple prose still resonates a hundred years later. Our scientific understanding—of atoms, life and the cosmos—has burgeoned in a fashion that not even he conceived: certainly Wells was right in predicting that the twentieth century would see more changes than the previous thousand years. Spinoffs from novel discoveries have transformed our world and our lives. The amazing technical innovations would surely have elated him, as would the prospects for the coming decades.

But Wells wasn't a naive optimist. His lecture highlighted the risk of global disaster: "It is impossible to show why certain things should not utterly destroy and end the human race and story; why night should not presently come down and make all our dreams and efforts vain . . . something from space, or pestilence, or some great disease of the atmosphere, some trailing cometary poison, some great emanation of vapour from the interior of the Earth, or new animals to prey on us, or some drug or wrecking madness in the mind of man." In his later years, Wells became more pessimistic, especially in his final book, *The Mind at the End of its Tether*. His near despair about the "downside" of science might have deepened were he writing today. Humans already have the wherewithal to destroy their civilisation by nuclear war: in the new century, they are acquiring biological expertise that could be equally lethal; our integrated society will become more vulnerable to cyber-risks; and human pressure on the environment is building up dangerously. The tensions between benign and damaging spinoffs from new discoveries, and the threats posed by the Promethean power science gives us, are disquietingly real, and sharpening up.

Wells's audience at the Royal Institution would have already known him as the author of *The Time Machine*. In this classic story the chrononaut gently eased the throttle of his machine forward: "night came like the turning out of a light, and in another moment came tomorrow." As he sped up "the palpitation of night and day merged into one continuous greyness. . . . I travelled, stopping ever and again, in great strides of a thousand years or more, drawn on by the mystery of the Earth's fate, watching with a strange fascination the sun grow larger and duller in the westward sky, and the life of the old Earth ebb away." He encounters an era where the human species has split into two: the effete and infantile Eloi, and the brutish underground Morlocks who exploit them. He ends up thirty million years hence, in a world where all familiar forms of life have become extinct. He then returns to the present, bringing strange plants as evidence of his trip.

In Wells's story it takes eight hundred thousand years for humans to divide into two subspecies, a time span that accords with modern ideas of how long it took for humanity to emerge via natural selection. (Evidence for our earliest hominoid ancestors extends back for four million years; it is about forty thousand years since "modern" humans superseded the Neanderthals.) But in the new century, changes in human bodies and brains won't be restricted to the pace of Darwinian selection, nor even to that of selective breeding. Genetic engineering and biotechnology, if widely practiced, could transmogrify humanity's physique and mentality far faster than Wells foresaw.

Indeed, Lee Silver, in his book *Remaking Eden*, conjectures that it could take only a few generations for humanity to divide into two species: if the technology enabling parents to "design" genetically advantaged children were available only to the wealthy, there would be a widening divergence between the "GenRich" and the "Naturals." Nongenetic changes could be even more sudden, transforming humanity's mental character in less than a generation, as quickly as new drugs can be developed and marketed. The fundamentals of humanity, essentially unaltered throughout recorded history, could start to be transformed within this century.

Failed Forecasts

I recently found in an antiquarian bookshop some science magazines, dating from the 1920s, with imaginative depictions of the future. The then-futuristic aeroplanes had rows of wings one above the other; the artist had surmised that since biplanes then seemed an advance on monoplanes, it would be still more "advanced" to stack wings like a Venetian blind. Extrapolation can be misleading. Moreover, straightforward projections of present trends will miss the most revolutionary innovations: the qualitatively new things that really change the world.

Even four hundred years ago, Francis Bacon emphasised that the most important advances are the least predictable. Three ancient discoveries especially astonished him: gunpowder, silk, and the mariner's compass. In *Novum Organum* he writes, "these things . . . were not discovered by philosophy or the arts of reason, but by chance and occasion," They are "different in kind," so that "no preconceived notion could possibly have conduced to their discovery." It was Bacon's belief that "there are still many things of excellent use stored up in the lap of nature having nothing in them kindred or parallel to what is already discovered . . . lying quite out of the path of imagination."

X rays, discovered in 1895, must have seemed fully as magical to Wells as the compass did to Bacon. Though of manifest benefit, they couldn't possibly have been planned for. A research proposal to make flesh appear transparent wouldn't have been funded, and even if it had been, the research surely wouldn't have led to the X ray. And the big discoveries have continued to take us unawares. Few managed to predict the inventions that transformed the world in the second half of the twentieth century. In 1937 the US National Academy of Sciences organised a study aimed at predicting breakthroughs; its report makes salutary reading for technological forecasters today. It came up with some wise assessments about agriculture, about synthetic gasoline, and synthetic rubber. But what is more remarkable is the things it missed. No nuclear energy, no antibiotics (though this was eight years after Alexander Fleming had discovered penicillin), no jet aircraft, no rocketry nor any use of space, no computers; certainly no transistors. The committee overlooked the technologies that actually dominated the second half of the twentieth century. Still less could they predict the social and political transformations that occurred during that time.

Scientists are often blind to the ramifications of even their own discoveries. Ernest Rutherford, the greatest nuclear physicist of his time, famously dismissed as "moonshine" the practical relevance of nuclear energy. The pioneers of radio regarded wireless transmission as a substitute for the telegraph, rather than as a means for "one-to-many" broadcasting. Neither the great computer designer and mathematician John von Neumann nor the IBM founder Thomas J. Watson envisaged a need for more than a few computing machines in the entire country. Today's ubiquitous mobile phones and palmtop computers would amaze anyone from a century ago; they are exemplars of Arthur C. Clarke's dictum that any sufficiently advanced technology is indistinguishable from magic. So what might happen in the new century that would be "magic" to us?

Forecasters have generally failed dismally to foresee the drastic changes brought about by completely unpredictable discoveries. In contrast, incremental change is often slower than forecasters expect, certainly far slower than is technically possible. Few have been as prescient as Clarke, but we will certainly have to wait until far later than 2001 before there are large space colonies or lunar bases. And the technology of civil aviation has stagnated, almost in the way that manned space flight has. We could have had hypersonic planes by now, but—basically for economic and environmental reasons—we don't: we cross the Atlantic in jets that have had basically similar performance for the last forty-five years, and are likely to for the next twenty. What has changed is the volume of traffic. Long-distance air travel has been transformed into an affordable mass market. Of course, there have been technical improvements, for instance in computerised control, and the precise positioning offered by global positioning system (GPS) satellites; for passengers the most conspicuous changes are in the sophistication of the gadgetry that provides on-board entertainment. Similarly, we drive cars that improve only incrementally over the decades. Transport technology in general has developed more slowly than many forecasters expected.

On the other hand, Clarke and most others were taken unawares by the speed with which personal computers proliferated and improved, and by spinoffs such as the Internet. The density with which circuits are etched on computer microchips has now been doubling every eighteen months for nearly thirty years in accordance with the famous "law" put forward by Gordon Moore, cofounder of Intel Corporation. In consequence, there is far more processing power in a computer-game console than was available to the Apollo astronauts when they landed on the Moon. My Cambridge colleague George Efstathiou, who simulates on a computer how galaxies form and evolve, can now repeat, on his laptop during his lunch break, calculations that took months on one of the world's fastest supercomputers then available when he first did them in 1980. Soon we will not merely have mobile phones, but high-bandwidth communication with everyone else, and instant access to all recorded knowledge. And the genomics revolution—a dominant feature of the early twenty-first century—is accelerating: when the great project to map the human genome began, few expected that it would be essentially completed by now.

Francis Bacon contrasted his three "magical" discoveries with the invention of printing, which "has nothing in it which is not open and generally obvious. . . when it had been made, it seems incredible that it should have escaped notice so long." Most inventions emerge, as printing did, by Bacon's second route: "from the transferring, composition, and application of [things] already known." The artefacts and gadgets familiar in everyday life are generally the outcome of a continuing trail of incremental improvement. But there can still be revolutionary innovations, despite the immense scientific infrastructure that was quite lacking in earlier centuries. Indeed, the lengthening frontiers of knowledge increase the chance of some remarkable surprises.

Faster Forward?

Over an entire century, we cannot set limits on what science can achieve, so we should leave our minds open, or at least ajar, to concepts that now seem on the wilder shores of speculative thought. Superhuman robots are widely predicted for mid-century. Even more astonishing advances could eventually stem from fundamentally new concepts in basic science that haven't yet even been envisioned and which we as yet have no vocabulary to describe. It is impossible to make firm projections that entail huge extrapolations of present knowledge

Ray Kurzweil, guru of "artificial intelligence" and author of *The Age of Spiritual Machines*, claims that the twenty-first century will see "20,000 years of progress at today's rate." That is just a rhetorical claim, of course, since "progress" can be quantified only within limited domains.

There are physical limits to how finely silicon microchips can be etched by present techniques, for the same reason that there are limits to the sharpness of the images that microscopes or telescopes can give us. But new methods are already being developed that can print circuits on a much finer scale, so "Moore's law" need not level off. Even within ten years, wristwatch-size computers will link us to an advanced internet and to the global positioning system. Looking further ahead, quite different techniques—tiny crisscrossing optical beams, not involving chip circuits at all—may increase computing power still further.

Miniaturisation, though already amazing, is very far indeed from its theoretical limits. Each tiny circuit-element of a silicon chip contains billions of atoms: such a circuit is exceedingly large and "coarse" compared to the smallest circuits that could in principle exist. These would have dimensions of only a nanometer—a billionth of a meter, rather than the micron (millionth of a meter) scale on which present-day chips are etched. One long-term hope is to assemble nanostructures and circuits "bottom up" by sticking single atoms and molecules together. This is how living organisms grow and develop. And it is how nature's "computers" are made: an insect's brain has about the same processing power as a powerful present-day computer.

The evangelists of nanotechnology envisage an "assembler" that could grab single atoms, shifting them around and assembling them one by one into machines with components no bigger than molecules. These techniques will allow computer processors to be a thousand times smaller, and information to be stored in memories a billion times more compact than the best we have today. Indeed, human brains may be augmented by implants of computers. Nanomachines could have as intricate a molecular structure as viruses and living cells, and display even more variety; they could carry out manufacturing tasks; they could crawl around inside our bodies observing and taking measurements, or even performing microsurgery.

Nanotechnology could extend Moore's law for up to thirty further years; by that time, computers would match the processing power of a human brain. And all human beings could by then be bathed in a cyberspace that allows instant communication with one another, not just in speech and vision but via elaborate virtual reality.

The robotics pioneer Hans Moravec believes that machines will attain human-level intelligence and may even "take over." For this to happen, processing power is not enough: the computers will need sensors that enable them to see and hear as well as we do, and the software to process and interpret what their sensors tell them. Advances in software have been far slower than in hardware: computers still can't match the facility of even a three-year-old child in recognising and manipulating solid objects. Perhaps more will be achieved by trying to "reverse-engineer" the human brain, rather than by just speeding up and compacting traditional processors. Once computers can observe and interpret their environment as adeptly as we do through our eyes and other sense organs, their far faster thinking and responses could give them an advantage over us. Then they will truly be perceived as intelligent beings, to which (or to whom) we can relate, at least in some respects, as we to other people. Ethical issues then arise. We generally accept an obligation to ensure that other human beings (and indeed at least some animal species) can fulfil their "natural" potential. Will we have the same duty to sophisticated robots, our own creations? Should we feel obligated to foster their welfare, and guilty if they are underemployed, frustrated, or bored?

A Human or Posthuman Future?

These projections assume that our descendants remain distinctively "human." But human character and physique will soon themselves be malleable. Implants into our brain (and perhaps new drugs as well) could vastly enhance some aspects of human intellectual powers: our logical or mathematical skills, and perhaps even our creativity. We may be able to "plug in" extra memory, or learn by direct input into the brain (the injection of an "instant Ph.D."?). John Sulston, a leader of the Human Genome Project, speculates on further implications: "How much non-biological hardware can we hook up to a human body and still call it human? ... A little more memory, perhaps? More processing power? Why not? And if so, perhaps a kind of immortality is just around the corner."

A further step would be to reverse-engineer human brains in enough detail to be able to download thoughts and memories into a machine, or reconstruct them artificially. Humans could then transcend biology by merging with computers, maybe losing their individuality and evolving into a common consciousness. If present technical trends proceeded unimpeded, then we should not dismiss Moravec's belief that some people now

living could attain immortality—in the sense of having a life
span that is not constrained by their present bodies. Those who
seek this kind of everlasting life will need to abandon their
bodies and have their brains downloaded into silicon hardware.
In old-style spiritualist parlance, they would "go over to the
other side."

A superintelligent machine could be the last invention that
humans need ever make. Once machines have surpassed human
intelligence, they could themselves design and assemble a new
generation of even more intelligent ones. This could then repeat itself, with technology racing towards a cusp, or "singularity," at which the rate of innovation runs away towards infinity.
(The Californian futurologist Vernor Vinge was the first to use
the term "singularity" in this apocalyptic context.) It is impossible to predict what the world might be like after the occurrence
of such a "singularity." Even the constraints based on currently
understood physical laws may be insecure. Some of the "sta-
ples" of speculative science that flummox physicists today—
time travel, space warps, and the like—may be harnessed by the
new machines, transforming the world physically as well.

Kurzweil and Vinge are of course on (or even beyond) the visionary fringe, where scientific prediction meets science fiction. Belief in the "singularity" relates to mainstream futurol-
ogy rather as the millenarian hope of "Rapture"—being
physically plucked up into the Heavens at an imminent Last
Day—relates to mainstream Christianity.

The Steady Backdrop

Information systems and biotechnology can surge ahead rapidly because (unlike, for instance, traditional forms of power
generation and transport infrastructure) they do not depend on

huge facilities that take years to construct and have to be operated for decades. But not everything is as mutable and transient as electronic hardware.

Barring some calamitous destruction—or unless there were indeed a technological surge towards a "singularity," after which superrobots could transform the world more drastically than we can now conceive—there are limits to how fast our terrestrial environment could alter. We will still have roads and (probably) railways, but these may be supplemented by novel means of travel (for example, GPS systems could allow automated collision-free journeys by land or air). The developing world could, on optimistic scenarios, acquire a new twenty-first-century infrastructure, unencumbered by the legacy of the past. But some limits are set by energy and resources: supersonic travel is unlikely to become routine for most of the world's population, unless some radically new plane design or engine is invented. Much travel will, however, become superfluous, superseded by telecommunication and virtual reality.

What about exploitation of space (perhaps using novel propulsion systems)? Robotics and miniaturisation are weakening the short-term practical case for manned space flight. In the coming decades, swarms of miniaturised satellites will orbit Earth; intricately instrumented unmanned probes will roam and explore throughout the solar system; and robotic fabricators will assemble large structures, perhaps extracting raw materials from the Moon or from asteroids. Within fifty years, if our civilisation escapes disastrous setbacks in the meantime, there could be a vibrant programme of human space exploration, though it is likely to be led by entrepreneurs and adventurers rather than by governments.

Even if there is an expanding human presence in space, it will involve only a trivial fraction of humanity. Nowhere away from

Earth offers a habitat that is even as clement as the Antarctic or the deep ocean bed; nonetheless, space may offer the backdrop for enthusiastic explorers and pioneers, who may eventually set up self-sustained social groups away from Earth. By the end of the century, such communities could have been established— on the Moon, on Mars, or freely floating in space—either as refuges, or in a spirit of exploration. Whether this happens, and how, could be crucial to posthuman evolution, and indeed to the fate of intelligent life in future centuries. Although it would be little consolation to those on Earth, life would have "tunnelled through" its era of maximum jeopardy: no terrestrial catastrophe could thereafter quench life's long-term cosmic potential.

The Real World: Longer Horizons

Techno-forecasters, their attitudes moulded by the social and political environment of the West Coast of the United States, where so many such people are congregated, tend to envisage that changes proceed untrammelled, in a social system supportive of innovations and that consumerist motivations dominate other ideologies. These presumptions may be as unwarranted as it would have been to downplay the role of religion in international affairs, or to predict that sub-Saharan Africa would have advanced steadily since the 1970s rather than regressing further into destitution. Unpredictable social and political developments add extra dimensions of uncertainty. Indeed, a main theme of this book is that technical advances will in themselves render society more vulnerable to disruption.

But even if disruption were no worse than it is today, these forecasts do little more than set the "envelope" of what might be possible: the gap between what is technically possible and what will actually happen is going to widen. Some innovations just don't attract enough economic or social demand: just as supersonic flight and manned space flight stagnated after the 1970s, today (in 2002) the potentialities of broadband (G3) technology are being taken up rather slowly because few people want to surf the Internet or watch movies from their mobile phones.

For biotechnologies, the inhibition will be more ethical than economic. If there were no regulations to rein back the application of genetic techniques, the physique and mentality of human beings could morph within a few generations. Futurists like Freeman Dyson speculate that within a few centuries, Homo sapiens may have diversified into numerous subspecies, adapting to a variety of habitats beyond Earth.

Economic decisions generally discount into insignificance what may happen more than twenty years from now: commercial ventures are not worthwhile unless they pay off far sooner than that, especially when obsolescence is rapid. Government decisions are often as short-term as the next election. But sometimes—in energy policy, for example—the horizon extends to fifty years. Some economists are trying to provide incentives for longer-term planning and prudent conservation by putting a monetary value on a country's natural resources, thereby rendering explicit in a nation's balance sheet the cost of depleting them. The debates about global warming that led to the Kyoto Protocol take cognisance of what might happen one or two centuries ahead: the consensus is that governments should take preemptive actions now, in the putative interest of our twenty-second-century descendants (though whether these actions will actually be implemented is still unclear).

There is one context in which official public policy looks even further ahead, not just for hundreds but for thousands of years: the disposal of radioactive waste from nuclear power stations. Some of this waste will remain toxic for many millennia;

both in the UK and the US, the specification for underground depositories demands that hazardous materials should remain sealed off—with no leakage via groundwater, or through fissures opened up by earthquakes—for at least ten thousand years. These geological requirements, imposed by the US Environmental Protection Agency, were important factors in the choice of a Nevada location, deep underground below Yucca Mountain, for the US's national waste dump.

The prolonged debates on radioactive waste disposal have had at least one benefit: they have generated interest and concern about how our present-day actions resonate through several millennia—time spans still infinitesimal, of course, compared to the future of Earth, but nonetheless far beyond the horizon of most other planners and decision-makers. The US Department of Energy even convened an interdisciplinary group of academics to discuss how best to design a message that could be understood by human beings (if any should exist) several millennia hence. Warnings unambiguous and universal enough to bridge any conceivable culture gap could be genuinely important in alerting our remote descendants to hidden dangers like radioactive waste depositories.

The Long Now Foundation, an initiative promoted by Danny Hillis (best known as inventor of the "Connection Machine," an early massively parallel processing computer), aims to promote long-term thinking by constructing a large ultra-durable clock that would record the passage of several millennia. Stewart Brand, in his book *The Clock of the Long Now*, discusses how to optimise the content of libraries, time capsules, and other enduring artefacts that could help to raise our gaze towards longer time horizons.

Even if changes proceed no faster than in the last few centuries, there will certainly be a "turnover" in cultures and political institutions within a single millennium. A catastrophic collapse of civilisation could destroy continuity, creating a gap as wide as the cultural chasm that we would now experience with a remote Amazonian tribe. In Walter M. Miller Jr.'s novel *A Canticle for Leibowitz*, North America reverts to a medieval state after a devastating nuclear war. The Catholic Church is the only institution to survive, and generations of priests attempt, for several centuries, to reconstruct prewar knowledge and technology from fragmentary records and relics. James Lovelock (best known as the originator of the "Gaia" concept, likening the biosphere to a self-regulating organism) urges compilation of a "start up manual for civilisation," copies of which should be dispersed widely enough to ensure that some survive almost any eventuality: it would describe techniques of agriculture, from selective breeding to modern genetics, and cover other technologies similarly.

By making us aware of longer time horizons, the proponents of the Long Now remind us that the welfare of far-future generations should not be jeopardised by imprudent policies today. But they are perhaps downplaying the qualitatively new consequences of computers and biotechnology. Optimists believe that these will lead to the transformations discussed in this chapter; realists accept that these advances will open up new peril. Prospects are so volatile that mankind might not even persist beyond a century—much less a millennium—unless all nations adopt low-risk and sustainable policies based on present technology. But that would require an infeasible brake on new discoveries and inventions. A more realistic forecast is that society's survival on Earth will, within this century, be exposed to new challenges so threatening that the radioactivity level in Nevada thousands of years from now will seem supremely irrelevant. Indeed, the next chapter suggests that we have been lucky to survive the last fifty years without catastrophe.

3

THE DOOMSDAY CLOCK

*Have We Been Lucky
to Survive This Long?*

---

*The Cold War exposed us to graver risks than
most would knowingly have accepted.
The danger of nuclear devastation still looms,
but threats stemming from new science are
even more intractable.*

THROUGHOUT MOST OF HUMAN HISTORY the worst
disasters have been inflicted by environmental forces—floods,
earthquakes, volcanoes and hurricanes—and by pestilence. But
the greatest catastrophes of the twentieth century were directly
induced by human agency: one estimate suggests that in the
two world wars and their aftermath, 187 million perished by
war, massacre, persecution, or policy-induced famine. The

twentieth century was perhaps the first during which more were killed by war and totalitarian regimes than by natural disasters. These man-made catastrophes were, however, played out against a backdrop of improving well-being, and not just in privileged countries, but in much of the developing world, where life expectancy at birth has almost doubled, and a smaller proportion live in abject poverty.

The second half of the twentieth century was beset by a menace far worse than any that had previously imperilled our species: the threat of all-out nuclear war. This threat has so far been averted, but it has hung over us for more than forty years. President Kennedy himself said during the Cuban Missile Crisis that the chance of nuclear war was "somewhere between one out of three and even." The risk was of course cumulative for several decades: at any time the response to a crisis could have escalated out of control; the superpowers could have stumbled towards Armageddon through muddle and miscalculation.

The Cuban missile standoff in 1962 was the event that brought us closest to a premeditated nuclear exchange. According to the historian Arthur Schlesinger Jr., one of Kennedy's aides at that time, "This was not only the most dangerous moment of the Cold War. It was the most dangerous moment in human history. Never before had two contending powers possessed between them the technical capacity to blow up the world. Fortunately, Kennedy and Khrushchev were leaders of restraint and sobriety; otherwise, we probably wouldn't be here today."

Robert McNamara was then the US secretary of defense, as he also was during the escalation of the Vietnam War. He later wrote that "Even a low probability of catastrophe is a high risk, and I don't think we should continue to accept it. . . . I believe that was the best-managed cold war crisis of any, but we came within a hairbreadth of nuclear war without realising it. It's no credit to us that we missed nuclear war—at least, we had to be lucky as well as wise. . . . It became very clear to me as a result of the Cuban missile crisis that the indefinite combination of human fallibility (which we can never get rid of) and nuclear weapons carries the very high probability of the destruction of nations."

We were all dragooned into this gamble throughout the Cold War era. Even the pessimists probably didn't rate the risk of nuclear war as being as high as fifty percent. So we shouldn't be surprised that we and our society survived; it was more likely that we would than that we wouldn't. Nonetheless, this does not necessarily mean that we were exposed to a prudent risk; nor does it vindicate the policy of the superpowers for several decades: nuclear deterrence by threat of massive retaliation.

Was the Risk Worth It?

Suppose you are invited to play Russian roulette (with one bullet in a pistol with six chambers) and told that if you survive, you will win fifty dollars. The most likely outcome (five to one in your favour) is that you will indeed end up better off: still alive, and with an extra fifty dollars in your pocket. Nonetheless, unless you hold your life very cheap indeed, this would be an imprudent—indeed blazingly foolish—gamble to have taken. The payoff would have to be very large before a sensible person would risk his or her life at these odds: many might be tempted if the potential prize were five million dollars rather than just fifty. Likewise, if you had a medical condition with a very poor prognosis without an operation, then—but only then—you might opt for surgery that carried a one in six chance of fatality.

So was it worth subjecting ourselves to the risks to which the entire planet was exposed during the Cold War? The answer depends, obviously, on what the probability of nuclear war actually was, something on which we can do no better than accept the views of officials like McNamara, who seems to rate it as having been substantially higher than one in six. But the answer also depends on our assessment of what would have happened without nuclear deterrence: how likely Soviet expansion would have been, and whether, in the words of the old slogan, you would "rather be red than dead." It would be interesting to know what risk the other leaders during that period actually believed they were exposing us to, and what risks most citizens would have accepted if they had been in a position to give informed consent. I personally would not have chosen to risk a one in six chance of a disaster that would have killed hundreds of millions and shattered the physical fabric of all our cities, even if the alternative was a certainty of a Soviet takeover of Western Europe. And of course the devastating consequences of nuclear war would have spread far beyond the countries that perceived that they were defending themselves against a genuine threat, and whose governments had implicitly taken this gamble: most of the Third World, already vulnerable to natural disasters, had this still greater hazard imposed on them.

A Science-Fuelled Arms Race

The *Bulletin of Atomic Scientists* was founded at the end of World War II by a group of physicists, based in Chicago, many of whom had worked at Los Alamos on the Manhattan Project, designing and building the atomic bombs dropped on Hiroshima and Nagasaki. It is still a thriving and influential journal, with a focus on arms control and nuclear policy. The "logo" on the cover of each issue is a clock, the closeness of whose hands to midnight indicates how precarious the world situation is—or is thought to be by the *Bulletin*'s editorial board. Every few years (sometimes more often) the minute hand is shifted, either forwards or backwards. These clock adjustments, stretching from 1947 to the present day, have tracked the successive crises in international relations: it is now closer to "midnight" than it was throughout the 1970s.

The era when the clock indicated maximal hazard was actually the 1950s: throughout that period it displayed a time of two or three minutes to midnight. In retrospect, this seems a correct judgement. Both the US and the Soviet Union acquired H-bombs during that decade, as well as larger numbers of atomic (fission) weapons. In retrospect, Europe was lucky to have escaped nuclear devastation in the 1950s. So-called battle-field nukes (one called the "Davy Crockett") were held at the battalion level; safeguards were less sophisticated than they later became, and there was real danger of a nuclear war starting by misjudgement or inadvertence; once triggered, it could have escalated out of control. The world seemed on a still shorter fuse when bombers were supplemented by much faster ballistic missiles that could cross the Atlantic within half an hour, allowing the other side only a few minutes to make the fateful choice whether to retaliate massively before their own arsenal was destroyed.

After the Cuban Missile Crisis, the nuclear danger rose higher on the political agenda: there was greater impetus towards arms control treaties, starting with a ban on nuclear tests in the atmosphere, signed in 1963. But there was no letup in the race to devise more "advanced" weaponry. McNamara noted that "virtually every technical innovation in the arms race has come from the US. But it has always been quickly matched by the other side." This syndrome was exemplified by the main development in the late 1960s. Engineers then devised how to carry multiple warheads on a single missile, and to aim them independently at different targets. This so-called MIRVing (the acronym stands for "multiple independently targeted reentry vehicle") was dreamed up by US technologists and then implemented both by them and by their Soviet counterparts. The net result of this, and other innovations, was to make both sides less secure. Each put the "worst case" construction on whatever the other side did, overestimated the threat, and overreacted.

Another innovation—antimissile missiles to protect cities and strategic sites against incoming warheads—was reined in by a bargain between the superpowers, the Antiballistic Missile (ABM) Treaty. Scientists helped to broker this agreement by behind-the-scenes arguments that any defence would destabilise the "balance of terror" and lead to countermeasures that would negate it.

In early 1980s the *Bulletin*'s clock was near midnight again. At that time, new medium-range nuclear weapons were introduced into the UK and Germany, allegedly to make more credible the threat of Western retaliation to a Soviet attack on Western Europe. The main issues were still how to reduce the ever-present risk of escalation towards catastrophic nuclear war, whether by malfunction, miscalculation, or premeditated strategy. The risk in a single year may have been small, but the probabilities would have multiplied if conditions had not changed.

The nuclear stockpile in the 1980s was equivalent to ten tons of TNT for each person in Russia, Europe, and America. Carl Sagan and others initiated a debate about whether an all-out nuclear exchange would trigger a nuclear winter: a worldwide blocking-out of the Sun, with results, including mass extinction, similar to those that would be triggered by the impact of a giant asteroid or comet. The eventual best guess was that not even the detonation of ten thousand megatons would have caused a prolonged worldwide blackout, though there are still uncertainties in the modelling (in particular, how high in the stratosphere the debris would reach, and how long it would stay there). But the "nuclear winter" scenario raised the disquieting prospect that the main victims of a nuclear war would be the populations of South Asia, Africa, and Latin America, mostly noncombatants in the Cold War.

This was the time of the Strategic Defence Initiative—"Star Wars"—which led to rearguing of the case for the Anti Ballistic Missile Treaty. It seemed technically impossible to construct a defensive "shield" effective enough to achieve President Reagan's proclaimed goal of making nuclear weapons "impotent and obsolete"; countermeasures always gave advantage to the offence. This treaty is now again under threat from the United States because it impedes the development of an antimissile defence system against putative missile launches from "rogue states." The main objection to this type of defensive system is that even if after vast expense and effort it worked, it would fail to counter the most basic nuclear threat from the "rogue states," the low-tech delivery of a bomb by ship or truck. Abrogation of the ABM treaty would also be regrettable because it would open the way to the "weaponisation" of space. Anti-satellite weapons are entirely feasible and would be relatively easy to develop. Compared to the challenge of intercepting an incoming missile, an object in a long-lived and predictable orbit would be a "sitting duck": communication, navigation, and surveillance satellites could easily be knocked out. Another risk is that a "rogue state" might be tempted to neutralise satellite-based antimissile defences by polluting space with orbiting debris, a stratagem that would stymie any use of space by low-orbiting satellites.

Solly Zuckerman, a long-time science adviser to the UK government, was (after his retirement) as eloquent as Robert McNamara in denouncing the dangerous absurdity of the chain of events that had built up the US and Soviet nuclear stockpiles to such a grotesque "overkill" level. According to Zuckerman, "The basic reason for the irrationality of the whole process [was] the fact that ideas for a new weapon system derived in the first place, not from the military, but from different groups of scientists and technologists. . . . A new future with its anxieties was shaped by technologists, not because they were concerned with any visionary picture of how the world should evolve, but because they were merely doing what they saw to be their job. . . . At base, the momentum of the arms race is undoubtedly fuelled by the technicians in governmental laboratories and in the industries which produce the armaments."

Workers in weapons laboratories whose skills rose above routine competence, or who displayed any originality, added their iota to this menacing trend. In Zuckerman's view the weapons scientists "have become the alchemists of our times, working in secret ways that cannot be divulged, casting spells which embrace us all. They may never have been in battle, they may never have experienced the devastation of war; but they know how to devise the means of destruction."

Zuckerman was writing in the 1980s. Further innovations would by now have ratcheted up the nuclear arms race by several more notches had not the agenda changed utterly. After the end of the Cold War the threat of a massive nuclear exchange no longer loomed so imminently over us (though thousands of missiles are still deployed by the US and Russia). In the early 1990s the *Bulletin*'s clock was put back to seventeen minutes to midnight. But it has been creeping forward again since then: in 2002 it was at seven minutes to midnight. We are confronted by proliferation of nuclear weapons (in India and

Pakistan, for instance), and by bewildering new risks and uncertainties. These may not threaten a sudden worldwide catastrophe—the Doomsday clock is not such a good metaphor—
but they are, in aggregate, as worrying and challenging. There
seems something almost comfortable, at least in retrospect,
about the paralytic but relatively predictable politics of Leonid
Brezhnev's "era of stagnation" and the superpower rivalry.

Huge nuclear stockpiles persisted throughout the 1990s, as
indeed they still do. Arms-control agreements to cut the number of deployed nuclear weapons are welcome, but they pose
the problem of managing and disposing of the twenty or thirty
thousand bombs and missiles still lying around. Treaties require that most of these warheads be dismantled. As an immediate measure, they can be put in a state of lower readiness or
alertness; targeting programmes can be countermanded; warheads can be taken out of missiles and stored separately. This
obviously puts everything on a longer fuse, and into a mode
where less manpower and expertise is needed to maintain the
arsenal safely. But it will take much longer—and be a major
technical challenge in itself—finally to get rid of all these
weapons, and to dispose safely of their uranium and plutonium.
Highly enriched uranium 235 can be rendered less dangerous,
though still usable in peaceful nuclear reactors, by mixing it
with uranium 238. In 1993 the US agreed to buy from Russia,
over a twenty-year period, up to five hundred tonnes of for-
merly weapons-grade uranium in this diluted form. Disposing
of plutonium is less straightforward. The Russians are reluctant to regard this hard-won material as "waste": however, existing nuclear power stations do not use the kind of "breeder"
reactors that can directly burn plutonium. The best options are
to bury it, or to render it unusable in weapons by mixing it with
radioactive waste or partially burning it in a nuclear reactor.
According to Richard Garwin and Georges Charpak, "The

total of excess material in Russia would provide something like 10,000 plutonium weapons and 60,000 uranium implosion weapons. Securing this material is truly a daunting task."

Until this disposal has been achieved, security and a reliable inventory must be maintained for all the weapons in the former Soviet Union: otherwise, far more could go astray than the entire stock of the "minor" nuclear powers. Indeed there is real disquiet—though no firm evidence—that during the transitional turbulence in the early 1990s, terrorist or rebel groups may already have purloined such weapons.

Construction of a long-range missile carrying a compact warhead is still far beyond the resources of dissident groups. But even this prospect has become less daunting and cannot be dismissed. For instance, now that the signals from GPS satellites are available to everyone, a cruise-type missile could be guided by a commercially available package. And a ground-hugging missile would be harder to track and intercept than a ballistic missile. Far less technically demanding techniques, which also would evade antimissile defences, include the detonation of a weapon transported in a truck or ship and the construction of a crude explosive device assembled, using stolen enriched uranium, in a city apartment. Unlike a missile-launched bomb, this would leave no trace of its provenance.

Countering the Spread of Weapons

In one respect, at least, the nuclear scene could be a lot worse. The number of nuclear powers has increased, but not as fast as many pundits had predicted. There may be up to ten, if one counts undeclared proliferators such as Israel; but at least twenty countries could have surmounted the technical threshold had they wished to but instead have eschewed any nuclear role: Japan, Germany, and Brazil, for instance. South Africa developed six nuclear weapons but has now dismantled them

When it began in 1967, the Nonproliferation Treaty (NPT) took cognisance of the special status of the five powers that already then possessed nuclear weapons: the US, the UK, France, Russia, and China. To make this "discrimination" less unpalatable to other nations, the treaty stated that these nuclear powers should "pursue negotiations in good faith on effective measures relating to cessation of the arms race . . . and the discontinuance of all test explosions of nuclear weapons for all time."

The NPT would have a fairer wind behind it if these five nations, as their side of the bargain, cut back their own arsenals more drastically. According to current treaties, it will take ten years before the US deployment falls even to two thousand warheads; moreover, the decommissioned warheads will not be irreversibly destroyed, but merely held in storage. The nuclear powers have also dragged their feet on a comprehensive test ban, which would curb the development of still more sophisticated weapons. The US has refused to ratify this treaty. Occasional testing is claimed to be needed to check that existing weapons in the stockpile remain "reliable"—in other words, that they would go off when they were supposed to. Debate continues about the extent to which reliability could be adequately assured by testing the components separately, by computer simulations, for example. It is in any case unclear how important this assurance is except for an aggressor planning a first strike: a nuclear missile remains a deterrent even if there is only a fifty percent chance its payload will explode. It is also claimed that tests are needed to ensure that the weapons are "safe"— that they will not explode or release dangerous radioactivity if accidentally mishandled. Another argument against a comprehensive test ban is that compliance cannot be adequately verified. Although underground tests of above a few kilotons have an unambiguous seismic signature, those below a kiloton may be drowned by the large number of small earthquakes, and can be muffled if they are carried out in large cavities. There is debate about how many seismic stations are needed for verification; and about how seismic evidence could be supplemented by intelligence, or by satellite surveillance. A report from the US National Academy of Sciences argues that undetectable tests would not be feasible, and tests are unnecessary to maintain existing stockpiles, only to develop new "advanced" weapons.

A comprehensive test ban would not in itself stop proliferation, because it is possible to make a credible first-generation fission bomb without a test. But a ban would inhibit the existing nuclear powers (particularly the US) from developing new types of bomb, and thereby improve the climate for the Nonproliferation Treaty, which enjoins all nuclear powers to reduce their arsenals. To counter proliferation, it is far more crucial to extend the role of the International Atomic Energy Agency in keeping track of fresh nuclear material and carrying out on-site inspections. This, of course, was the issue that triggered the crisis over Iraq.

But the most important determinant will be whether nations perceive an incentive to join the nuclear club. The existing nuclear powers could help by downplaying the role of nuclear arms in their defence postures. Recent statements by the US, and even the UK, on the possible use of low-yield nuclear weapons to attack underground hideouts are in this regard a real step backwards. Such declarations blur the nuclear threshold, and make the use of nuclear weapons less unthinkable; they increase the incentive for other countries to get their own bombs, an incentive that is already strengthening because there seems no other way to deter or counter unwelcome pressure

from the US, whose advantage in "smart" conventional weaponry is so overwhelming that the superpower can impose its
will on other nations with minimal human cost to itself.

Concerned Scientists

The Chicago atomic scientists were not the only ones who attempted from outside government to influence the political debate about the post–World War II nuclear threat. Another
group founded a series of conferences that took the name of
the village Pugwash, in Nova Scotia, where the first such conference was held under the sponsorship of a Canadian million-
aire, Cyrus Eaton, who had been born there. The participants
in early Pugwash conferences came from the Soviet Union as
well as the West, and had generally been active in World War
II; they had worked on the bomb project, or on radar, and had
maintained an informed concern ever since. Particularly during
the 1960s and 1970s, the Pugwash conferences provided valuable informal contact between the US and the Soviet Union
when there were few formal channels.

There are still some remarkable survivors from this generation. The most senior is Hans Bethe, born in 1906 in Stras-
bourg, in Alsace–Lorraine. In the 1930s he was already eminent as a nuclear physicist. He moved from Germany to an
academic post in the US and during World War II became
head of the theoretical division at Los Alamos. He afterwards
returned to Cornell University, where even in the new century
he has continued to be active in promoting arms control, as
well as pursuing research (his main recent interest being in the
theory of exploding stars and supernovae). Bethe must rank as
the most universally respected of all living physicists, acclaimed

not only for his science but for his sustained concern and involvement with its implications. He is perhaps unique among physicists in having published fine work for more than seventy-five years. In 1999 his attitude towards military research hardened, and he urged scientists to "cease and desist from work creating, developing, improving, and manufacturing nuclear weapons and other weapons of potential mass destruction" on the grounds that this fuelled the arms race.

Another Los Alamos veteran whom I have been privileged to know is Joseph Rotblat. Two years younger than Bethe, he experienced in his Polish childhood the hardships of World War I, and began his career as a research scientist in his home country. In 1939 he came as a refugee to England to work with the eminent nuclear physicist James Chadwick at Liverpool; his wife was never able to join him, and she perished in a concentration camp. Rotblat joined the Manhattan project at Los Alamos as part of the small British contingent. But he chose to leave prematurely when it became clear that German defeat was near, because in his mind the bomb project could be justified only as a counterbalance to a possible nuclear weapon in Hitler's hands. Indeed, he recalls having been disillusioned by hearing General Groves, head of the project, saying as early as March 1944 that the main purpose of the bomb was "to subdue the Russians."

Rotblat returned to England, where he became a professor of medical physics, doing pioneering research into the effects of exposure to radiation. In 1955 he encouraged Bertrand Russell to prepare a manifesto stressing the urgency of reducing the nuclear peril. One of Einstein's last acts was to agree to be a cosignatory. This eloquent manifesto, whose authors claimed to be "speaking on this occasion not as members of this or that nation, continent or creed, but as human beings, members of the species Man, whose continued existence is in doubt," led to the initiation of the Pugwash conferences in 1957; ever since, Rotblat has been their "prime mover" and untiring inspiration. When the achievements of these conferences were recognised by the 1995 Nobel Peace Prize, it was fitting that half the award went to the Pugwash organisation and half to Rotblat personally. Rotblat, now aged 94, still pursues, with the dynamism of a man half his age, his unflagging campaign to rid the world completely of nuclear weapons. This is often derided as an unrealistic goal, espoused only by fringe groups and idealists of a woolly and unthinking kind. Rotblat remains an idealist, but without illusions about the gap between hope and expectation, and his cause is broadening its support.

"The proposition that nuclear weapons can be retained in perpetuity and never used—accidentally or by decision—defies credibility." This firm declaration comes from a 1997 report of an international group convened by the Australian government and known as the Canberra Commission. Its members included not only Rotblat but also Michel Rocard, former prime minister of France; Robert McNamara; and retired military and air force generals. The commission noted that the only military utility of nuclear weapons was to deter their use by others, and put forward step-by-step proposals for moving, in a politically stable way, towards a world without nuclear weapons.

Those who were uprooted from placid academic laboratories to join the Manhattan project belonged to what seems in retrospect the "golden generation" of physicists: many had been pivotal in establishing our modern view of atoms and nuclei. They were mindful that fate had plunged them into epochal events. Most of them returned to academic work in universities, but sustained a lifelong concern with nuclear weapons. All were deeply marked by their involvement, but in divergent ways, as exemplified by the contrasting postwar careers of the two most prominent personalities, J. Robert Oppenheimer and Edward Teller. (Andrei Sakharov, the most celebrated Soviet counterpart of these two Americans, came from a slightly younger generation, having been involved in the postwar development of the H-bomb.)

The Chicago atomic scientists, and the pioneers of the Pugwash movement, set an admirable example for researchers in any branch of science that has grave societal impact. They did not say that they were "just scientists" and that the use made of their work was up to politicians. They took the line that scientists have a duty to alert the public to the implications of their work, and should retain a concern with how their ideas are applied. We feel there is something lacking in parents who are unconcerned about what happens to their children in adulthood, even though it is generally beyond their control. Likewise, scientists should not be indifferent to the fruits of their research: they should welcome (and indeed try to foster) benign spin-offs, but resist, so far as they can, dangerous or threatening applications.

In the present century the dilemmas and threats will come from biology and computer science, as well as from physics: in all these fields society will insistently need latter-day counterparts of Bethe and Rotblat. University scientists and independent entrepreneurs have a special obligation because they have more freedom than those in government service and company employees subject to commercial pressures.

4

POST–2000 THREATS
Terror and Error

---

*Within twenty years, bioterror or bioerror could
kill a million people. What does this presage
for later decades?*

I AM FINALISING THIS CHAPTER IN DECEMBER 2002, just
over a year after the September 11 attacks on the United
States. There is continuing fear that further outrages will in-
scribe other tragic dates in our collective memories. A succession of suicide bombers is terrorising Israel. The bombers are
intelligent young Palestinians (women as well as men) with
warped idealism. In the late twentieth century, organised terrorists groups with rational political aims (for instance, those
operating in Ireland) refrained from the very worst they could
have done because, even with their distorted perspective, they
reckoned that beyond a certain threshold an outrage would be

counterproductive to their cause. The Al Qaeda terrorists who crashed the planes on the World Trade Center and the Pentagon had no such inhibitions. If such groups were to obtain a nuclear weapon, they would willingly detonate it in a city centre, killing tens of thousands along with themselves; and millions around the world would acclaim them as heroes. The consequences could be even more catastrophic if a suicidal zealot were to become intentionally infected with smallpox and trigger an epidemic; in future there could be viruses even more lethal (and without an antidote).

The Einstein–Russell manifesto had this to say about the concerns of well-informed scientists in the 1950s with regard to the nuclear threat: "None of them will say that the worst results are certain. What they do say is that these results are possible, and no one can be sure that they will not be realised. We have not yet found that the views of experts on these questions depend in any degree on their politics or prejudices. They depend only, so far as our researches have revealed, on the extent of the particular expert's knowledge. We have found that the [experts] who know most are the most gloomy."

The same could be said today about other risks that now loom just as large. Twenty-first-century technology confronts us with a diverse array of lethal prospects that were not yet on the horizon during the Cold War era. Moreover, the potential perpetrators are also more diverse, and more elusive. The prime new threats are "asymmetric": they come not from nation states but from subnational groups, and even from individuals.

Even if all nations impose strict regulations on the handling of nuclear material and dangerous viruses, the chances of effective enforcement, worldwide, are no better than current enforcement of laws against illegal drugs. Just one infringement could trigger widespread disaster. Such risks plainly can never be completely eliminated. But far worse, they seem set to become more intractable and threatening. There will always be disaffected loners in every country, and the "leverage" that each can exert is increasing. And there are other quite different threats. In cyberspace, for instance, there is a race between attempts to render systems more robust and secure, and the growing ingenuity of criminals who may try to infiltrate and sabotage those systems.

Nuclear Megaterror

Nuclear "megaterrorism" is a prime risk. Tom Clancy's novel *The Sum of Our Fears*, turned into a film released in 2002, portrayed devastation of a crowded football stadium by a purloined nuclear device. Nuclear energy is a million times more efficient, per kilogram, than chemical explosions. The bomb used in the Oklahoma City attack, which killed over 160 people—until September 11, 2001, the worst-ever attack on the US homeland—was equivalent to about three tonnes of TNT. The nuclear stockpiles of the former Soviet Union and the US amount to that much explosive power for each person in the world, and hence the danger if even a minuscule fraction of this arsenal—even a single one of the tens of thousands of warheads that now exist—were to go astray.

Nuclear bombs fuelled by plutonium have to be triggered by a precisely configured implosion. This is technically challenging, perhaps too challenging for terrorist groups. But plutonium could be coated on the surface of a conventional bomb to make a "dirty bomb." Such a weapon would cause no more immediate fatalities than a large conventional bomb, but would create extensive long-term disruption because it would pollute a large area with unacceptable levels of radiation. A still greater terrorist risk comes from enriched uranium (separated U-235) because it is far easier to make a genuine nuclear explosion using this fuel. The Nobel physicist Luis Alvarez claimed, "With modern weapons-grade uranium. . . terrorists would have a good chance of setting off a high-yield explosion simply by dropping one half of the material onto the other half. Most people seem unaware that if separated U-235 is to hand, it's a trivial job to set off a nuclear explosion, whereas if only plutonium is available, making it explode is the most difficult technical job I know." Alvarez is unduly downplaying the difficulty of making a uranium weapon. However, an explosion could be achieved by using a cannon or mortar to propel a subcritical mass, configured as a shell or bullet, into another subcritical mass shaped into a ring or hollow cylinder.

A nuclear explosion at the World Trade Center, involving two grapefruit-sized lumps of enriched uranium, would have devastated three square miles of southern Manhattan, including the whole of Wall Street. It would have killed hundreds of thousands if it went off during working hours. Similar devastation would arise if there were attacks on other cities. And conventional explosives could trigger disaster on almost the same scale if, for example, they were set off so as to detonate huge storage tanks of oil or natural gas. (Indeed, the 1993 bombing of the World Trade Center could have been as destructive as that of 2001 if the explosion, set off at one corner of the foundations, had caused one tower to topple and crash onto the other tower.)

"We have slain the dragon, but are now living in a jungle full of poisonous snakes," said James Wolsey, former director of the CIA, in 1990. He was referring to the turbulence that followed the collapse of the Soviet Union and the end of the Cold War. A decade later, his metaphor is even more appropriate for the elusive groups that threaten us.

These short-term risks highlight the urgency of safeguarding the plutonium and enriched uranium in the republics of the former Soviet Union. It could be already too late. Stewardship was lax in the political turmoil of the early 1990s: Chechen rebels and other subnational groups may already have appropriated some weapons.

In 2001 the US cut back on a proposed three-billion-dollar subvention to Russia and the other states of the former Soviet Union for decommissioning weapons, preventing "defection" of scientific experts, and disposing of plutonium—efforts that surely deserve far more urgent priority than "national missile defence." A positive development, however, has been the Nuclear Threat Initiative, chaired by ex-Senator Sam Nunn and funded primarily by Ted Turner, founder of CNN, which is using its own resources and political leverage to energise threat-reduction measures.

Terrorism is a new risk affecting our attitude to civilian nuclear power stations—augmenting the traditional liabilities of high capital cost, decommissioning problems, and the legacy of toxic waste left for future generations. A power station harbours not only the highly radioactive "core," but also a stock of spent fuel-rods that could be more vulnerable. Even the latter, if set on fire, could release ten times more cesium-137 (with a thirty-year half-life) than the Chernobyl accident.

Designers of nuclear reactors aimed to reduce the probability of the worst accidents to less than one per million "reactor years." To do such calculations, all possible combinations of mishaps and subsystem failures have to be included. Among these is the possibility that a large aircraft might crash onto the containment vessel. Air accident records (and projections into the future) tell us how many aircraft are likely to fall out of the sky. In the whole of Europe and North America it is only a few per year. The chance that one of them would hit a particular building is reassuringly low, much less than one in a million per year. But we now know that this is not the right calculation. It overlooks the possibility, now nightmarishly familiar, that kamikaze-style terrorists could aim for just such a target, using a large fully fuelled jet, or a smaller plane loaded with explosives. The chance of such an event cannot be assessed even by the most astute technicians or engineers: it is a matter of political or sociological judgement. But one would surely have to be a naive optimist to rate it as less than one in a hundred per year. If this high estimate had been fed into risk assessments when nuclear power stations were being planned, then current designs might not have been sanctioned. It could become incumbent on all new designs to meet safety standards that may even require them to be put underground.

The role of nuclear power could in any case decline during the next twenty years if existing nuclear power stations reach the end of their lives and are not replaced. Many thousands of new power stations would be needed if nuclear energy were to contribute substantially to the worldwide goal of reducing greenhouse emissions. Quite apart from the sabotage and terrorism threats, the risk of accidents rises if maintenance is lax. The poor safety records of some Third-World airlines endanger primarily those who fly in them; poorly maintained reactors pose a threat that doesn't respect national boundaries.

Nuclear power could have a brighter future if novel kinds of fission reactors that overcome the safety and decommissioning problems of present designs came into routine use. Another long-range prospect is nuclear fusion: a controlled version of the process that keeps the Sun shining and powers the H-bomb. Fusion has long been touted as an inexhaustible source of energy. But the goal has receded: after a false dawn back in the 1950s before the real difficulties were realised, fusion has consistently seemed at least thirty years away.

The prime advantage of nuclear power, whether fusion or fission, is that it simultaneously solves two problems: limited oil reserves and global warming. But a preferable option, on both environmental and security grounds, would be renewable sources. These will surely supply an increasing fraction of the world's needs, but won't be able to supply total demand without some technical breakthroughs. Wind turbines alone won't be enough, and current solar energy conversion is too expensive and inefficient. But if sunlight could be harnessed by some cheap and effective photovoltaic material that can be draped over huge areas of unproductive land, then the so called "hydrogen economy" would be feasible: solar-generated electric power would extract hydrogen from water; this hydrogen can then be used in fuel cells, which substitute for internal combustion engines.

Biothreats

More disquieting than nuclear dangers are the potential hazards stemming from microbiology and genetics. For decades several nations have had substantial and largely secret programmes to develop chemical and biological weapons. There is ever-growing expertise in designing and dispersing lethal pathogens, not least in the US and UK, where there are continuing research programmes to improve countermeasures against biological attacks. Iraq is suspected of pursuing an offensive programme; several other countries (South Africa, for instance) have had such programmes in the past.

Back in the 1970s and 1980s the Soviet Union was engaged in the largest-ever mobilisation of scientific expertise to develop biological and chemical weapons. Kanatjan Alibekov was at one time the number-two scientist in the Soviet *Biopreparat* programme; he defected to the US in 1992, Westernising his name to Ken Alibek. According to his book *Biohazard*, he was in charge of more than thirty thousand workers. He recounts the efforts made to modify organisms to make them more virulent and more resistant to vaccines. In 1992, Boris Yeltsin admitted something that Western observers had long suspected: at least 66 mysterious deaths in the city of Sverdlovsk that occurred in 1979 were caused by anthrax spores that had leaked from a *Biopreparat* laboratory.

The problem of detecting illicit fabrication of nuclear weapons is as nothing compared with the task of verifying national compliance with treaties on chemical and biological weapons. And even that is easy compared with the challenge of monitoring subnational groups and individuals. Biological and chemical warfare were long regarded as cheap options for states without nuclear weapons. But it no longer requires a state, or even a large organisation, to mount a catastrophic attack: the resources needed could be acquired by private individuals. The manufacture of lethal chemicals or toxins requires modest-scale equipment that is, moreover, essentially the same as is needed for medical or agricultural programmes: the techniques and expertise are "dual use." This is another contrast with nuclear programmes, where the uranium enrichment needed for efficient fission weapons requires elaborate equipment with no legitimate alternative use. In the words of Fred Ikle, "The knowledge and techniques for making biological superweapons will become dispersed among hospital laboratories, agricultural research institutes, and peaceful factories everywhere. Only an oppressive police state could assure total government control over such novel tools for mass destruction."

Thousands of individuals, perhaps even millions, may someday acquire the capability to disseminate "weapons" that could cause widespread (even worldwide) epidemics. A few adherents of a death-seeking cult, or even a single embittered individual, could unleash an attack. Indeed, there have already been small-scale bioattacks, but fortunately, the techniques were too primitive, and too ineptly executed, to accomplish even as much as a conventional explosive could have done. In 1984 some followers of the Rajneeshee cult (he of the yellow robes and fifty Rolls Royces) contaminated some salad bars in Wasco County, Oregon, with salmonella, and 750 people were stricken with gastroenteritis. The motive for the attack was apparently to incapacitate votes in a local election, and thereby influence the outcome of a planning application for the cult's commune. But the origin of this epidemic was inferred only a year later, highlighting the problem of tracing the perpetrators of any biological attack. In the early 1990s the Aum Shinrikyo sect in Japan developed various agents including botulinum toxin, Q fever, and anthrax. They released the nerve gas sarin in the Tokyo subway, killing twelve; the attack could have been far more devastating had they been more successful in dispersing the gas in the air.

In September 2001, envelopes containing anthrax spores were sent to two US senators and to several media organisations. Five people died—a tragedy, but on a scale no larger than everyday road crashes. However—and this is an important portent—the blanket media coverage in the US led to a "dread factor" that pervaded the entire nation. One can readily envisage the massive consequence for the national psyche of an outrage that killed thousands. The actual impact of a future attack could be greater if an antibiotic-resistant variant of the bacterium were used, and, of course, if it were dispersed effectively. This threat is leading to a biological "arms race": efforts to develop drugs and viruses that can target specific bacteria, and also sensors to detect pathogens in very low concentrations.

What Could a Bioattack Currently Achieve?

Many studies and exercises have been undertaken in order to gauge the possible impact of a bioattack and how emergency services might respond to it. Back in 1970 the World Health Organisation estimated that a release of fifty kilograms of anthrax spores from an aeroplane flying upwind of a city could cause nearly one hundred thousand deaths. More recently, in 1999, several scenarios were explored by the Jason group, a consortium of highly rated academic scientists who do regular consulting for the US Department of Defense. The group considered what would happen if anthrax were released in the New York subway. Spores would be dispersed along the tunnel system and by passengers. If the release had been covert, the first evidence would emerge a few days later when the victims (by then widely spread over the country) visited their doctors with symptoms.

The Jason group also studied the effects of a chemical agent, ricin, which attacks ribosomes and interferes with the chemistry of proteins. A lethal dose is only ten micrograms. However, the fact that the sarin attack on the Tokyo subway did not kill thousands showed that the spread of the agent is not a trivial technical challenge. Details have been released of experiments on (nontoxic) aerosol dispersals carried out in the 1950s and 1960s in the US and the UK. These were done in the London underground, the New York subway, and in San Francisco.

Achieving efficient dispersal in the air is a generic problem with all chemical agents, as it is with biological agents (like anthrax) that are not infectious. To say that a few grams of an agent could in principle kill millions may be true, but it may also be misleading (just as it would be misleading to say that one man could father a hundred million children; spermatozoa are plentiful enough, but dispersal and delivery would be a real challenge).

For infectious diseases, initial dispersal is less crucial than for anthrax (which cannot be passed on from person to person); even a localised release, especially in a mobile population, could trigger a widespread epidemic. Perhaps the most fearful prospect, among known viruses, is smallpox. Through a magnificent worldwide effort in the 1970s, spearheaded by the World Health Organisation, the disease has been completely eradicated. Rather than making the virus extinct, stocks have been maintained in two locations, the Center for Disease Control, in Atlanta, USA, and the ominously named Vector Laboratory in Moscow. The justification for preserving these viruses is that they could be used to help develop vaccines. However, there is growing concern that clandestine caches of the virus may exist in other countries, raising fears of smallpox bioterror.

Smallpox is highly contagious (almost as infectious as measles) and kills about one-third of those who succumb to it. There are several published studies of what would happen if this deadly virus were released. Even if the epidemic were contained, and casualties ran only into hundreds, the effect on a large city could be devastating. There would be a run on medical supplies, especially if vaccines were scarce. But an actual death toll could run into millions, especially if the epidemic spread internationally.

In July 2001 an exercise, entitled "Dark Winter," simulated a covert smallpox attack on the United States and the response and countermeasures to it. In this exercise, roles were played by experienced figures: the former US senator Sam Nunn played the president, and the governor of Oklahoma played himself. It was assumed that aerosol clouds contaminated with smallpox virus were released simultaneously in three locations—shopping malls—in different states. The scenario led, in the worst case, to three million people being infected (of whom a third would have died). Prompt vaccination would eventually have choked off the spread of the disease (the vaccine is still effective even four days after infection occurs). But an infection that spread worldwide, as it would if the initial release were at an airport or in an airliner, could start a runaway epidemic in countries where vaccine was not so readily available as in the US—worst of all, perhaps, in the congested megacities of the developing world. The incubation period is twelve days, so by the time the first case was manifest, those originally infected would have spread around the world, and induced secondary infections. It would be too late to impose any effective quarantine.

"Smallpox 2002: Silent weapon," a docudrama broadcast by the BBC, portrayed a single suicidal fanatic in New York who infected enough people to trigger a pandemic that claimed sixty million victims. This scarifying scenario was based on a (perhaps questionable) computer model of how the virus would spread. When mathematicians try to compute how an epidemic develops, the most crucial factor that enters their calculations is the number of people a typical victim infects, known as the "multiplier." For this particular model, this number was assumed to be 10. But some experts have argued that smallpox is not so infectious, that it typically takes several hours of proximity to pass on and that these scenarios therefore exaggerate the ease with which an infected person transmits the disease. However, there is evidence (for instance, from a 1970 outbreak in a German hospital) that the virus can be spread by air currents as well as physical contact. Some experts have suggested that a multiplier of 10 may be appropriate in hospitals, but only 5 in the community: others suggest that the multiplier could actually be as low as 2.

Uncertainties like this are crucial in determining how readily an epidemic could be contained by mass vaccination or quarantine. But of course, it would be harder to control an outbreak if (as envisaged in the BBC scenario) it had spread, before being detected, into developing countries where reaction to such an emergency would be slower and less effective. And there will surely be other viruses that are still more readily transmitted. In the UK an epidemic of foot and mouth disease in 2001 had disastrous countrywide consequences for agriculture despite maximal efforts to control it. The outcome would be far worse if such an infection were maliciously spread. Bioattacks threaten people and animals; but they could threaten crops and ecosystems as well. Another of the Jason group's near-term scenarios was an attempt to sabotage agricultural production in the American Midwest by introducing the fungus known as "wheat rust," a naturally occurring fungus that sometimes destroys up to ten percent of the crop in California.

One feature common to all biological attacks is that they cannot be detected until it is too late, perhaps even not even before the effects have diffused worldwide. Indeed, the use of bioweapons in organised warfare has been inhibited not only be moral reservations, but because the time lag and spread cannot be controlled by military commanders. But this delay is an attraction to the lone dissident or terrorist, because the provenance of an attack—when or where the pathogen was released—can be readily camouflaged. The prospects of early detection would be improved by rapid nationwide sharing and analysis of medical information so that it would be easier to spot a sudden rise in the number of patients displaying a specific set of symptoms, or the near-simultaneous incidence of some rare or anomalous syndrome.

Any attack would induce severe disruption and panic. The alarmist reporting of the 2001 anthrax episode in the US exemplified how even a localised threat can affect the mindset of a whole continent. By amplifying fears and fuelling hysteria, media coverage would guarantee that even a smallpox epidemic at the less severe end of the spectrum of predictions would disrupt ordinary life worldwide.

Engineered Viruses?

All pre-2000 epidemics (with the possible exception of the 1979 Russian anthrax release) were caused by naturally occurring pathogens. But the biothreat has been aggravated by the advance of biotechnology. According to a report issued in June 2002 by the US National Academy of Sciences, "Just a few individuals with specialised skills and access to a laboratory could inexpensively and easily produce a panoply of lethal biological weapons that might seriously threaten the US population. Moreover, they could manufacture such biological agents with commercially available equipment—that is, equipment that could also be used to make chemical, pharmaceuticals, foods, or beer—and therefore remain inconspicuous. The deciphering of the human genome sequence and the complete elucidation of numerous pathogen genomes . . . allow science to be misused to create new agents of mass destruction."

The report notes that the new technology should also, as an "upside," lead to quicker ways of identifying and reacting to a pathogen release, but its overall message is disquieting. It recognises that a skilled "loner" could perpetrate a catastrophic epidemic, even though the focus is now on terrorist groups. All over the world there are people with the expertise to undertake genetic manipulations and cultivate microorganisms. George Poste, a British biotechnologist and government advisor who now works in the US, conjectures that "It would be interesting to reflect, if [the "Unabomber"] had been trained in the 1990s, whether he would have chosen to use bombs or would have walked along and dropped something into a hamburger plant as an alternative, as the ubiquity of 'Biotechnology 101' becomes increasingly commonplace in the university curriculum around the world." (In 2002 the US approved a huge increase in funding for biodefence. This will, as an unwelcome byproduct, disseminate such expertise more widely.)

Eckard Wimmer and his colleagues at the State University of New York announced in July 2002 that they had assembled a polio virus, using DNA and a genetic blueprint that could be downloaded from the Internet. This artificial virus posed little hazard, because most people have been immunised against polio. But it would be no more difficult to create variants that could be infectious and even lethal. Experts had known for years that the kind of synthesis Wimmer did was feasible; some criticised him for doing an unnecessary experiment just as a stunt. But to Wimmer it was a "scary realisation" that viruses could be so readily created. Viruses like smallpox, with larger genomes than the polio virus, pose a greater technical challenge; moreover, the smallpox virus would not be able to reproduce itself unless replication enzymes were added from a different pox virus. However, some smaller and equally lethal viruses—HIV and ebola, for instance—could be created, even now, by assembling a chromosome from individual genes, as Wimmer did.

Within a few years, the genetic blueprints of vast numbers of viruses, as well as of animals and plants, will be archived in laboratory databases accessible to other scientists via the Internet. The blueprint of the ebola virus, for example, is already archived; there are thousands of people with the skills to assemble it, using strands of DNA that are available commercially. In the 1990s, members of the Aum Shinrikyo cult tried to track down the naturally occurring ebola virus in Africa: it is fortunately rare, and they failed to find it. They would now find it easier to assemble it in a home laboratory. Home computers and the Internet have opened up immensely greater scope for amateur scientists. In a subject like astronomy, this is an important and unreservedly welcome development. But one would view ambivalently the empowerment of a sophisticated community of amateur biotechnologists.

Creation of "designer viruses" is a burgeoning technology.

And a better understanding of the human immune system, though of crucial medical benefit, will also make it easier for those who wish to suppress immunity. A succession of different engineered viruses for which there was no immunity nor antidote could have an even more catastrophic effect worldwide than AIDS is now having in Africa (where it is reversing decades of economic advance): for instance an equivalent of smallpox for which there is no vaccine, perhaps even a virus that spreads even more readily than smallpox itself, or a variant of AIDS that is transmitted like influenza, or a version of ebola with a longer gestation period. (Outbreaks of this dreadful contagious disease are usually contained because it acts so fast, killing its victims by eroding away their flesh before they have much chance to infect others. In contrast, it is the slowness with which AIDS acts that allows it to be effectively transmitted.)

Unless the capability to design new viruses is matched by corresponding skills in designing and manufacturing vaccines targeted against them, we could find ourselves as vulnerable as the native Americans who succumbed to diseases brought by European settlers, to which they had no immunity.

Strains of bacteria can be developed that are immune to antibiotics. Indeed, such bacteria are already emerging naturally as an outcome of Darwinian selection. Some hospital wards have already been infected by "bugs" resistant even to vancomycin, the antibiotic of last resort. Artificial engineering may "ring the changes" more effectively than natural mutations. New organisms could be designed to attack plants and even inorganic substances.

We may not have to wait long before new kinds of synthetic microbes are being genetically engineered. Craig Venter, former chief executive officer of Celera, the company that sequenced the human genome, has already announced plans to help solve the world's energy and global warming crises by creating new microbes: one type would dissociate water into oxygen and hydrogen (for the "hydrogen economy"); others would feed off carbon dioxide in the atmosphere (thereby combating the greenhouse effect) and convert it into organic chemicals of the kind that are now made from oil and gas. Venter's technique involves making an artificial chromosome with about five hundred genes and inserting it into an existing microbe whose own genome has been destroyed by radiation. If this technique works, it opens up the prospects of designing new forms of life that could feed off other materials in our environment. For example, fungi could be designed that could feed on and destroy polyurethane plastics. Even machines could be under threat: specially designed bacteria could change oil into a crystalline material, and thereby clog up machinery.

Laboratory Errors

Almost as worrying are the growing risks stemming from error and the unpredictable outcomes of experiments, rather than from malign intent. A recent episode in Australia was a worrying portent. Ron Jackson was a researcher at the Animal Control Cooperative Research Centre in Canberra, a government laboratory whose main mission was to improve techniques for controlling animal pests. With his colleague Ian Ramshaw he was researching new ways to bring down the population of mice. Their idea was to modify the mousepox virus so that it became, in effect, an infectious contraceptive vaccine, and use it to sterilise mice. During these experiments, early in 2001, they inadvertently created a new, highly virulent, strain of mousepox: their laboratory mice all died. They had added a gene for a protein (interleukin–4) that boosted antibody production and suppressed the immune system in the mice; in consequence, even animals that had previously been vaccinated against mousepox died as well. Had these scientists been working instead on the smallpox virus, could they have modified it to become even more virulent, so that vaccination offered no protection against it? According to Richard Preston, "the main thing that stands between the human species and the creation of a supervirus is a sense of responsibility among the individual biologists."

This type of laboratory experiment, creating pathogens more dangerous than predicted, and perhaps more virulent than have ever developed naturally, is an exemplar of a kind of hazard that scientists will need to confront (and seek to minimise) in other research areas. These areas include nanotechnology (and even fundamental physics), where the consequences could be even more calamitous. Nanotechnology holds great long-term promise, but could eventually have an even more serious downside than any bioerror. It is conceivable—though still far from reality—that nanomachines could be devised that can assemble copies of themselves. If let loose, their numbers could rise exponentially, until they ran out of "food." If their consumption were highly selective, these machines could be useful substitutes for chemical factories, in the same way as "designer bugs" could be. But the danger arises if nanomachines could be designed to be more omnivorous than any bacterium, perhaps even able to consume all organic materials. Metabolising efficiently, and utilising solar energy, they could then proliferate uncontrollably, and not reach the Malthusian limit until they had consumed all life.

This chain of events is dubbed by Eric Drexler the "grey goo scenario." He writes, "'Plants' with 'leaves' no more efficient than today's solar cells could out-compete real plants, crowding the biosphere with an inedible foliage. Tough omnivorous 'bacteria' could out-compete real bacteria. They could spread like blowing pollen, replicate swiftly and reduce the biosphere to dust in a matter of days. Dangerous replicators could easily be too tough, small, and rapidly spreading to stop—at least if we make no preparations. We have trouble enough controlling viruses and fruit flies."

The resultant population explosion in these "biovorous replicators" could then in theory devastate a continent within a few days. This is very much a theoretical "worst case"; nonetheless, such estimates carry the message that if the technology of self-replicating machines were ever developed, a fast-spreading disaster could not be ruled out.

Can the "grey goo" threat be taken seriously, even if we extend our forecast a century ahead? A runaway plague of these replicators would not violate basic scientific laws. But that does not make it a serious risk. To take another futuristic technology: an antimatter-powered space rocket attaining ninety percent of the speed of light is compatible with basic physical laws, but we know that it is technically far beyond us. Perhaps these hyperefficient replicators, feeding off the biosphere, are just as unrealistic as a "star ship," another example of where the "envelope" of what is consistent with general scientific laws (and hence theoretically possible) is far above what is likely. Should we categorise the ideas of Drexler, and others as scare-mongering science fiction?

Viruses and bacteria are themselves superbly engineered nanomachines, and an omnivorous eater that could thrive anywhere would be a winner in the natural selection stakes. So if this plague of destructive organisms is possible, Drexler's critics might argue, why didn't it evolve by natural selection, long ago? Why didn't the biosphere self-destruct "naturally," rather than being threatened only when creatures designed by misapplied human intelligence are let loose? A riposte to this argument is that human beings are able to engineer some modifications that nature cannot achieve: geneticists can make monkeys or corn glow in the dark by transferring a gene from a jellyfish, whereas natural selection cannot bridge the species barriers in this way. Likewise, nanotechnology may achieve in a few decades things that nature never could.

After 2020, sophisticated manipulations of viruses and cells will become commonplace; integrated computer networks will have taken over many aspects of our lives. Any forecasts for the mid-century are the realm of conjectures and "scenarios." By then, nanobots could be a reality; indeed, so many people may try to make nanoreplicators that the chance of one attempt triggering disaster would become substantial. It is easier to conceive of extra threats than of effective antidotes.

Such seemingly remote concerns should not divert attention from the diverse vulnerabilities described in this chapter that are already with us, and growing. The prospects should make us at least as "gloomy" as the pioneer atomic scientists were, half a century ago, when the nuclear threat emerged. The gravity of a threat is its magnitude multiplied by its probability: this is how we estimate our concern about hurricanes, asteroid impacts, and epidemics. If we apply this calculus to the future man-made risks that confront us, adding them all together, the Doomsday Clock will shift ever closer to midnight.

5

PERPETRATORS AND PALLIATIVES

When just a few technically adept individuals can threaten human society, abandonment of privacy may be the minimal price for maintaining security. But would even a "transparent society" be safe enough?

WE ARE ENTERING AN ERA when a single person can, by one clandestine act, cause millions of deaths or render a city uninhabitable for years, and when a malfunction in cyberspace can cause havoc worldwide to a significant segment of the economy: air transport, power generation, or the financial system. Indeed, disaster could be caused by someone who is merely incompetent rather than malign.

These threats are growing for three reasons. First, the destructive and disruptive capabilities available to an individual trained in genetics, bacteriology, or computer networks will grow as science advances; second, society is becoming more integrated and interdependent (internationally as well as nationally); and third, instant communications mean that the psychological impact of even a local disaster has worldwide repercussions on attitudes and behaviour.

The most conspicuous sub-national threat today comes from Islamic extremists, motivated by traditional values and beliefs far removed from those prevailing in the US and Europe. Other causes and grievances, also pursued rationally and single-mindedly, can inspire equally fanatical acts by sectarian groups or even "loners." Moreover there are some—and their numbers may grow in the US—who have a tenuous hold on rationality, and could pose an even more intractable threat if they had access to advancing technology.

Techno-Irrationality

Some optimists imagine that scientific or technical education reduces the propensity towards extreme irrationality and delinquency. But many examples belie this. The "Heaven's Gate" cult, though small in scale, was a portent of what can happen in the technocratic West. A "cell" of cult members in California formed an enclosed community, adept enough to finance themselves by designing web pages for the Internet. But their technical competence, and genuine interest in space technology and other sciences, was allied with a belief system that defied the rationality of scientific thought. Many cult members had actually castrated themselves: they proclaimed on their web site the aspiration to transmogrify themselves into "a physical body belonging to the true Kingdom of God—the Evolutionary Level above human—leaving behind this temporal and perishable world for one that is lasting and non-corruptible."

The arrival of the beings who they believed would transport them to this higher plane would be heralded by a comet: "Comet Hale–Bopp's approach is the 'marker' we've been waiting for—the time for the arrival of the spacecraft from the Level Above Human to take us home to Their World. We are happily prepared to leave This World." When this comet, one of the brightest of the last decade, approached closest to Earth, thirty-nine cult members, including their leader Marshall Applewhite, aseptically and methodically took their lives.

Of course, collective suicides are nothing new: they date back at least two thousand years. And they continue into modern times even in the west. The Reverend James Jones led a messianic cult that retreated to a remote location in South America—"Jonestown," in Guyana. In 1978 he instigated a mass suicide that left all nine hundred cult members dead of cyanide poisoning.

Although modern technology allows instant worldwide communication, it actually makes it easier to survive within an intellectual cocoon. The Heaven's Gate group didn't need to go to the Amazonian forest to be isolated: being economically self-sufficient via the Internet they could cut themselves off from any contact with their actual physical neighbours, indeed with any "normal" people. Instead, their beliefs were reinforced by selective electronic contact with other adherents of the cult on other continents.

The Internet offers access, in principle, to an unprecedented variety of opinions and information. Nonetheless, it could narrow understanding and sympathies rather than broaden them: some people may choose to stay closeted within a cybercommunity of the likeminded. In his book *republic.com*, Cass

Sunstein, a law professor at the University of Chicago, suggests that the Internet is allowing all of us to "filter" our input, so that each person reads a "Daily Me" customised to individual tastes and (more insidiously) purged of material that may challenge prejudices. Rather than sharing experience with those whose attitudes and tastes are different, many will in future "live in echo chambers of their own design" and "need not come across topics and views that you have not sought out. Without any difficulty, you are able to see exactly what you want to see, no more and no less." It is too early to predict the effect of the Internet on mainstream society (particularly in an international context). But there is a danger that it promotes isolation and allows us (if we choose) more easily to evade the everyday contacts that would unavoidably bring us up against conflicting views. Sunstein discusses "group polarization," whereby those who interact only with the likeminded have their prejudices and obsessions reinforced, and shift towards more extreme positions.

The creed of "Heaven's Gate" was an amalgam of "new age" and science fiction concepts. This cult was not unique; indeed, it is perhaps part of a resurgent trend. The Raelians, headquartered in Canada, have over fifty thousand adherents in more than eighty countries. Their founder and leader, Claude Vorilhon, originally a motor-racing journalist, claimed in 1973 to have been abducted by aliens and given information about how the human race was created using "DNA technology." The Raelians are aggressively promoting a programme of human cloning that is not only ethically problematic, but seems dangerously premature even to advocates of this technique.

These cults may seem to come from the same "fringes" as UFO-watching conspiracy theorists and the like. But in the US equally bizarre beliefs seem almost part of the mainstream. Millions believe in "the Rapture"—when Christ swoops back to Earth and transports true believers up to Heaven—or in the imminent Millennium, as depicted in the Book of Revelation. The long-term future of this planet and its biosphere are of no consequence to these Millenarian believers, some of whom have achieved influence in the US. (During the Reagan administration, environmental and energy policies were entrusted to James Watt, a religious fundamentalist who held the office of secretary of the interior. He believed that the world would end before the oil was used up and before we suffered the consequences of global warming or deforestation, so it was almost our duty to be profligate with Earth's divinely provided resources.)

Some of these cultists, like the members of "Heaven's Gate," threaten only themselves. It would be unfair to demonize them all, or to conflate very disparate beliefs. The resurgent cults are of course still a minuscule "sideshow" compared with traditional ideologies. The zealotry of traditional religious enthusiasts, allied to the single-issue fanaticism and ruthlessness of (for instance) the animal rights extremists in the US and the UK, can be a threatening mix, especially when accompanied by technical sophistication. The Internet allows groups to organise as well as offering access to technical expertise. Our social and economic system is becoming so brittle and interconnected that just a few individuals with this mindset and access to modern technology can exert huge "leverage."

Even if one disruptive event could be coped with, a succession of them, their psychological impact amplified by ever more pervasive communications media, would be cumulatively corrosive. Awareness that such events could occur without warning would exact a heavy social cost. In terrorist-prone localities you are reluctant to venture on a bus if you fear there may be a suicide bomber among your fellow passengers; you hesitate to do favours for a stranger; the privileged seek shelter in gated communities and enclaves. Future mega-terror could engender, worldwide, this breakdown of community and trust.

Obviously, these concerns offer a further incentive for nations and the international community to minimise the disaffection and injustices that give pretexts for grievance. But it is clear even from recent US experience that the internal problem of nihilistic or apocalyptic cults and aggrieved individuals is an intractable one.

Is Intrusive Surveillance the Least-Bad Safeguard?

One palliative would be acceptance of a complete loss of privacy, with the deployment of novel techniques to keep tabs on us all. Universal surveillance is becoming technically feasible, and could plainly be a safeguard against unwelcome clandestine activities. Techniques such as surgically implanted transmitters are already being seriously mooted to (for instance) monitor criminals on parole. Subjecting all citizens to such treatment would be deeply unpalatable to most of us, but if the threats escalated, we might become resigned to the need for such measures, and the next generation might find it less repugnant.

Orwellian surveillance, in traditional totalitarian style, would plainly be unacceptable; unless encryption techniques kept pace, it would become ever more intrusive with each technical advance. But suppose the surveillance was two-way, and each of us could "spy" not only on the government but on everyone else. The science fiction writer David Brin, in *The Transparent Society*, perhaps provocatively, has argued that this "symmetrical" (but even more intrusive) surveillance might be the least unacceptable way to ensure a safer future. It would obviously require a change in mindset. But this may come about. Closed circuit TV systems in public places are widespread in Britain, and generally welcomed as reassuring security measures, despite the loss of privacy. More and more information about us—what we buy, where and when we travel, and so forth—is already being recorded on "smart cards" used to purchase merchandise or travel tickets, and whenever we use a mobile phone. I am surprised how many of my friends willingly put intimate personal matters on web pages, open to the world. So a "transparent society," in which deviant behaviour couldn't escape notice, may be accepted by its members in preference to the alternatives.

Futuristic scenarios conjured up in Europe and the US may seem only marginally relevant to the rest of the world, where poverty deprives most people of the basic benefits of twentieth-century technology. But this transparency could spread worldwide, as mobile phones and the Internet have done.

How would this spread affect relations between rich and poor nations? Few non-Africans have direct knowledge of sub-Saharan Africa except through films and TV news reports. How will the US and European perceptions of the rest of the world change when there can be immediate personal links? An optimistic view would be that graphic "real time" evidence of personal need—of, for instance, the AIDS sufferers who cannot afford even a dollar a day for basic treatment—would stimulate individual generosity more effectively than the occasional messages and photographs received by contributors to traditional programmes of charitable sponsorship. But it seems unlikely that those who in the US retreat to gated communities, insulated from the poor even in their own neighbourhoods, would reach out to the desperate people of Africa. Even if they had the chance to befriend them and maintain video contact, "compassion fatigue" could quickly set in. Indeed, this may be another instance where the cyberworld leads to sharper social segmentation.

On the other hand, those in Africa and South Asia will have their relative deprivation brought more insistently to their consciousness, especially if (as is possible) access to cyberspace becomes cheaper to provide than basic sanitation, food, and healthcare. The millions in impoverished countries will become less quiescent, more aware of contrasts with more privileged areas, and with the technical means to create major disruption. It is not just religious fundamentalism that can trigger angry hostility to the West. If the entire developing world adopted so-called Western values, the disadvantaged would be even more embittered at the unequal benefits from globalisation and at a system of economic incentives that provides superfluities for the rich rather than necessities for the destitute.

Can We Stay Human?

Up until now, it has been religion, ideology, culture, economics, and geopolitics that have moulded societies. All these elements—in their immense variety—have been the pretext for internal disputes and for wars. One unchanging element over the centuries, however, has been human nature. But in the twenty-first century drugs, genetic modification, and perhaps silicon implants into the brain will change human beings themselves—their minds and attitudes, even their physiques.

Future genetically induced changes in the human population—though vastly faster than any naturally occurring evolutionary changes—would still require a few generations. But alterations in mood and mentality could spread even more rapidly through entire populations via addictive drugs (or perhaps by electronic implants).

In *Our Posthuman Future* Francis Fukuyama argues that habitual and universal use of mood-altering medications would

narrow and impoverish the range of human character. He cites
the use of Prozac to counter depression, and of Ritalin to damp
down hyperactivity in high-spirited but otherwise healthy children: these practices are already constricting the range of personality types that are considered normal and acceptable.
Fukuyama foresees a further narrowing, when other drugs are
developed, that could threaten what he regards as the essence
of our humanity.

Indeed, injection of hormones that act directly on the brain
will soon be able to effect far more powerful and "targeted"
changes in our personality than Prozac and its ilk. The hor-
mone PYY 3–36 has been shown to eliminate feelings of
hunger, by acting directly within the hypothalamus. One of the
specialists in this technique, Steve Bloom, of Hammersmith
Hospital, London, has expressed concerns about where this
work could lead even within ten years: "If we can alter people's
desire for food, we can alter other deep-seated desires: the hypothalamus is also home to brain circuits that influence sex
drive and sexual orientation."

Fukuyama fears that drugs will become universally used to
tone down extremes of mood and behaviour, and that our
species could degenerate into pallid acquiescent zombies: society would become a dystopia resembling Aldous Huxley's *Brave
New World*. Even if we looked the same, we wouldn't be fully
human. Fukuyama would favour strong control of all mind-al-
tering drugs. Prohibitions need not be one hundred percent effective if the aim was to stave off the day when all extreme personalities could be erased. There would be little overall impact
on national character if, despite regulation, a few delinquents
gained access to drugs by illicit tactics, or by travelling from
their own country to one with laxer regulations.

But my worry is the obverse of Fukuyama's. "Human nature"
encompasses a rich variety of personality types, but these in-

clude those who are drawn towards the disaffected fringe. The destabilising and destructive influence of just a few such people will be ever more devastating as their technical powers and expertise grow, and as the world we share becomes more interconnected.

Thirty years ago the psychologist B. F. Skinner foresaw, in his book *Beyond Freedom and Dignity*, that some form of mind control might be needed to avoid a breakdown of society; he argued that "conditioning" of the entire population was a prerequisite for a society that its members were content to live in and that none wished to destabilise.

Skinner was a behaviourist, and his mechanistic "stimulus-response" theories are now discredited. But the issue that he highlighted is now more acute because scientific advances allow even a single "aberrant" personality to cause widespread havoc. If a present-day psychologist were emboldened to offer a panacea, it would, ironically, resemble Fukuyama's posthuman nightmare: a population rendered docile and law-abiding by "designer drugs" and genetic intervention that can "correct" extremes of personality. Future brain science may even be able to "modify" the personalities of people whose mindset might lead them to become dangerously disaffected: an even more dystopian prospect.

In Philip K. Dick's science fiction fantasy *Minority Report* (now a Steven Spielberg movie) the "pre cogs," mentally abnormal human beings specially bred for the role, can identify those who are likely to commit future crime; potential felons are then pre-emptively tracked down and imprisoned in vats. If our propensities are indeed determined by genetics and physiology (and it is still unclear to what extent they are) then identifying potential criminals may soon not require psychic powers. There will then be growing pressures to institute this kind of pre-emptive action in the real world, as a safeguard against the

outrages—ever more calamitous with each technical advance—
that could be wrought even by one delinquent individual.

Our civilisation, as Stewart Brand notes, is "ever more
tightly linked and ever more leveraged out over the abyss on an
elaborate superstructure of highly sophisticated technology,
every part of which depends on the success of every other
part." Can its essence be safeguarded, without humanity having
to sacrifice its diversity and individualism? Must we, to survive,
be cowed by a police state, deprived of all privacy, or tranquil-
lised into passivity?

Or could the threats be reduced by putting the brakes on
potentially threatening science and technology and even
renouncing some areas of scientific research completely?

6

SLOWING SCIENCE DOWN?

Twenty-first century sciences offer bright prospects, but have a dark side as well. Ethical constraints on research, or relinquishment of potentially threatening technologies, are difficult to agree and even tougher to implement.

IN 2002 THE MAGAZINE *WIRED*, a glossy monthly with a focus on computers and electronic gadgetry, inaugurated a series of "long bets." The idea was to gather some predictions about future developments in society, science, and technology, and thereby stimulate debate. The Internet guru Esther Dyson forecast that Russia would achieve supremacy in the world's software industry within ten years. Physicists staked bets on how long it would take to formulate a unified theory of the fundamental forces—and indeed, on whether such a theory even exists. Another bet was that someone now alive would live to the age of 150, which is not implausible, given the rate of medical advance, but an odd bet insofar as the prognosticators would not themselves expect to survive long enough to witness the outcome.

I staked one thousand dollars on a bet: "That by the year 2020 an instance of bioerror or bioterror will have killed a million people."

Of course, I fervently hope to lose this bet. But I honestly do not expect to. This forecast involved looking less than twenty years ahead. I believe the risk would be high even if there were a "freeze" on new developments, and the potential perpetrators of such outrages or megaerrors had continuing access only to present-day techniques. But of course, no subject is forging ahead faster than biotechnology, and its advances will intensify the risks and enhance their variety.

Anxiety within the scientific community seems surprisingly muted. New technologies can plainly offer colossal benefits, and most scientists have the attitude that the downsides can often be best remedied by more (or differently directed) technology; they are mindful of how much we might forgo if we did not press ahead. In the early days of steam, hundreds of people died horribly when poorly designed boilers exploded; likewise, aviation was hazardous in its early days. Most surgical procedures, even if now routine, were risky and often fatal when they were being pioneered. Every advance has proceeded by "trial and error," but the acceptable threshold can be set higher when the risk is voluntarily accepted and the possible "upside" is large (as in the case of surgery). In an essay entitled "The hidden cost of saying no," Freeman Dyson highlighted this issue. He emphasised that the development and introduction of new drugs is inhibited—sometimes to the detriment of many whose lives could be saved thereby—by the prolonged and expensive safety trials required before approval.

But there is a difference when those exposed to the risk are given no choice, and don't stand to gain any compensating benefit, when the "worst case" could be disastrous, or when the risk can't be quantified. Some scientists seem fatalistic about the risks; or else optimistic, even complacent, that the more scarifying "downsides" can be averted. This optimism may be misplaced, and we should therefore ask, can the more intractable risks be staved off by "going slow" in some areas, or by sacrificing some of science's traditional openness?

Scientists accept the need for controls on the way they work, and how their discoveries are applied. Biological advances are opening up an ever increasing number of potential applications—human cloning, genetically modified organisms, and the rest—where regulation will be called for. Almost any applicable discovery has a potential for evil as well as for good. No responsible scientist would echo the words of H.G. Wells's fiendish Dr. Moreau: "I went on with this research just the way it led me. That is the only way I ever heard of true research going. I asked a question, devised some method of getting an answer, and got a fresh question. . . . The thing before you is no longer an animal, a fellow-creature, but a problem. . . . I wanted . . . to find out the extreme limit of plasticity in a living shape."

Scientific Self-Restraint

Restraint is obviously justified if the experiments themselves pose a risk: for instance, by creating dangerous pathogens that might escape, or by generating extreme concentrations of energy. Scientists sometimes abide by self-imposed moratoriums on specific lines of research. A precedent for this was the declaration put forward in 1975 by prominent molecular biologists to refrain from some types of experiments rendered possible by the then-new technique of recombinant DNA. This followed a meeting at Asilomar, California, convened by Paul Berg, of Stanford University. The Asilomar moratorium soon came to seem unduly cautious, but that doesn't mean that it was unwise at the time, since the level of risk was then genuinely uncertain. James Watson, codiscoverer of DNA's double helix, regards this attempt at self-regulation as, in retrospect, a mistake. (Watson is generally "bullish" about the applications of biotechnology, believing that we should be uninhibited about using new knowledge of genetics to "improve" humanity. He has asked rhetorically "If biologists won't play God, who will?"). But another Asilomar participant, David Baltimore, remains proud of the episode: in his view it was right "to engage society in thinking about the problems, because we know that society could block us from realising the tremendous benefits of this work unless we square with them and lead them in thinking through the problems."

The Asilomar episode seemed an encouraging precedent. It showed that an international group of leading scientists could agree to a self-denying ordinance, and that their influence on the research community was sufficient to ensure that it was implemented. There are now even more reasons for exercising restraint, but a voluntary consensus would be far harder to achieve today: the community is far larger, and competition (enhanced by commercial pressures) is more intense.

In many countries there are formal guidelines and licensing requirements for animal experimentation, motivated by humane concerns. However, there is a "penumbra" of experiments that, even though neither cruel nor dangerous, prompt a reflex of revulsion that leads some to urge broader regulations.

Bioethicists use the term "yuck factor" to denote an emotional recoil from violations of what we perceive as the natural order. This reaction sometimes just reflects an unthinking conservatism that erodes as we become familiar with a new technique: kidney transplants provoked this response when first introduced but are now widely accepted; indeed, even cornea transplants once did. Newspaper pictures of a mouse that had been implanted with a template on which tissue had grown in the form of a human ear, almost as large as the rest of its body, prompted an exaggerated "yuck!" reaction, despite assurances that the mouse was itself relaxed about its treatment and oblivious to the way it looked.

I personally have a "yuck!" response to invasive experiments that modify how animals behave. Physiologists at the State University of New York's medical centre in Brooklyn implanted electrodes into rats' brains. One electrode stimulated the cerebral "pleasure centre"; two other electrodes activated the regions which process signals from its left and right whiskers. This simple procedure transformed the animals into "roborats" that could be steered to left or right, and forced to behave in patterns that would seem to go squarely against their instincts. These procedures were not necessarily cruel to the rats, and in some sense no different from the way a horse or ox is harnessed or driven. Nonetheless, such experiments could presage intrusive modifications (of humans as well as of animals) that undercut what many feel should be their intrinsic nature; the same reaction will be engendered by more sophisticated hormonal techniques for modifying thought processes.

Perhaps only a minority react in this disproportionate way against these mouse and rat experiments. However, some procedures that may soon be possible, could trigger such widespread revulsion that there will surely be pressure to ban them: for instance, the "design" of insentient animals who (it would then be argued) would have the moral status of vegetables and so could be treated appallingly with no ethical compunctions at all. (The food industry would then be relieved of pressure to abandon its cruelly intensive treatment of factory-farmed animals). Brainless hominoids whose organs could be harvested as spare parts would seem, ethically, even more problematic. On the other hand, transplanting organs from pigs or other animals into humans should raise no more ethical concerns than meat-eating, though this technique (xenotransplantation) will perhaps be banned—irrespective of ethical judgements—because of the risks that new animal diseases could be introduced into the human population. Using stem cells to grow a replacement organ in situ would seem much the most acceptable alternative to transplant surgery, which often involves waiting tensely, with ambivalence if not eagerness, for a car crash or similar misfortune to provide a suitable "donor."

Animal cloning techniques may soon become routine, but attempts to clone human beings provoke a widespread "yuck!" response. The Raelian cult is rumoured already to have hundreds of cloned embryos. Responsible scientists would oppose any cloning attempts because of the likelihood that even if a pregnancy proceeded to full term, the resulting infant could be devastatingly damaged. Despite the general ethical objections, and the high chance of defective births, it is surely only a matter of years before the birth of the first cloned human.

Choices on how science is applied—to medicine, the environment, and so forth—should be debated far beyond the scientific community. This is one reason why it is important that a wide public should have a basic feel for science, knowing at least the difference between a proton and a protein. Otherwise, such debate won't get beyond slogans, or will be conducted at megaphone level via sensational headlines in tabloid newspapers. The views of scientists should not have special weight in deciding questions that involve ethics or risks: indeed, such judgements are best left to broader and more dispassionate groups. One welcome feature of the publicly funded Human

Genome Project was that part of the budget was specifically allocated to discussion and analysis of the project's ethical and societal impact.

The Paymasters of Science

Scientific research, and our motives for pursuing it, cannot be separated from the social context in which such research is carried out. Science underpins modern society. But equally, society's attitudes determine what kind of science is found interesting, and what projects gain favour with governments or commercial patrons.

There are several examples just from sciences I am myself involved in. Huge machines for studying subatomic particles gained government funding because they were spearheaded by physicists who had achieved clout through their role in World War II. The sensors used by astronomers to detect faint emission from distant stars and planets were devised to enable the US military to detect Vietnamese in the jungle; they are now used in digital cameras. And expensive scientific projects in space—the probes that have landed on Mars and given close-up pictures of Jupiter and Saturn—ride along on a huge space programme that was initially driven by superpower rivalry during the Cold War. The Hubble Space Telescope would have cost even more had it not shared some development costs with spy satellites.

Because of extraneous influences like this—and one could come up with equivalent lists in other scientific fields—scientific effort is deployed suboptimally. This seems so whether we judge in purely intellectual terms, or take account of likely benefit to human welfare. Some subjects have had the "inside track" and garnered disproportionate resources. Others, such as environmental research, renewable energy sources, and bio-diversity studies, deserve more effort. Within medical research the focus is disproportionately on cancer and cardiovascular studies, the ailments that loom largest in prosperous countries, rather than on diseases endemic in the tropics.

Most scientists nonetheless regard knowledge and under-standing as worth attaining for their own sake, believing that "pure" research should be untrammelled, provided that it is safe and there are no ethical objections. But is that simplistic? Are there areas of academic research—the kind of science done in university laboratories—that the wider public should try to restrain, because of unease about where they might lead? The surest safeguard against a new danger would be to deny the world the basic science that underpins it.

All countries give enhanced support, on strategic grounds, to sciences that promise valuable spin-off. (Molecular biology is favoured over black hole research, for example; I am myself in-volved with the latter, but nonetheless this discrimination doesn't seem unfair to me.) But does the converse follow: should support be withdrawn from a line of "pure" research, even if it is undeniably interesting, if there is reason to expect that the outcome will be misused? I think it should, especially since the present allocation among different sciences is itself the outcome of a complicated "tension" between extraneous fac-tors. Of course, scientists cannot be completely stopped from thinking and speculating: their best ideas often come unbidden, during leisure hours. But any academic scientist whose grant has been stopped is aware that funding cuts can slow down a line of research, even if they could never halt it completely,

Whenever an investigation holds short-term promise of lu-crative spin-off, public funding isn't needed, because commer-cial sources will step in to bankroll it: only government regula-tion could then stop such research. Such regulation would also constrain how private benefactors choose to dispense their resources. Wealthy individuals can distort research—one American gave Texas A&M University five million dollars for research on cloning because he wanted to clone his elderly dog.

To put effective brakes on a field of research would require international consensus. If one country imposed regulations, the most dynamic researchers and enterprising companies would migrate to another that was more sympathetic or permissive. This is happening already in stem cell research, where some countries, particularly the UK and Denmark, have established relatively permissive guidelines, and are thereby attracting a "brain gain." By offering a still more enticing regime to researchers and to their fledgling biotech industry, Singapore and China aim to leapfrog the competition.

The difficulty with a dirigiste policy in science is that the epochal advances are unpredictable. I already noted that X rays were an accidental discovery by a physicist, not the outcome of a crash medical program to see through flesh. As another example, a nineteenth-century project to improve reproduction of music would have led to an elaborate and mechanically intricate orchestrion, but would have brought us no closer to the techniques actually used in the twentieth century. These techniques were the outcome of curiosity-driven research on electricity and magnetism by Michael Faraday and his successors. In more recent times, the pioneers of lasers had little concept of how their invention would be applied (and certainly did not expect that one of the first uses would be for operations to repair detached retinas).

We can ask of any innovation whether its potential is so scary that we should be inhibited in pressing on with it, or at least impose some constraints. Nanotechnology, for instance, is likely to transform medicine, computers, surveillance, and other practical areas, but it might advance to a stage at which a replicator, with its associated dangers, became technically feasible. There would then be the risk, as there now is with biotechnology, of a catastrophic "release" (or that the technique could be used as a "suicide" weapon); the only countermeasure would be a nanotech analogue of an immune system. To guard against this, Robert Freitas suggests an Asilomar-style moratorium: artificial life should be studied only via computer experiments, rather than by experimenting with any kind of "real" machines, and there should be a ban on developing nanomachines that can reproduce in a natural environment. Similar concerns might be raised about superintelligent computer networks and other extrapolations of present technology.

Concealment or Openness?

Rather than aiming to slow down a research area, can the risks be stanched by selectively denying new knowledge to those who seem likely to misapply it? Governments have always kept much of their defence-related work secret. But research that is not so classified (nor kept confidential for commercial reasons) has traditionally been made accessible to everyone. In 2002 the US government proposed that scientists themselves should restrict the dissemination of new research that, though not classified, is sensitive and could be misapplied: this was such a departure from the usual ethos that it caused controversy within the American scientific community.

What does a university do if a seemingly qualified student with a lavish grant but suspicious provenance wants to enrol for a Ph.D. in nuclear engineering or microbiology? By attempting to obstruct the training of potential delinquents we could at most impose a modest delay in the diffusion of new ideas, especially since "high-risk" individuals cannot be reliably identified anyway. Some may say that anything that applies the brakes, even marginally, is worthwhile. Others might argue that since the capability will spread anyway, it might even be better to be networked to as many ex-students as possible. There is then less chance that a substantial illicit project could be pursued without news leaking out via personal contacts. Maximum openness in communications, and a high rate of international migration, would render even small-scale clandestine projects harder to conceal. The international flow of students and scholars is in practice restricted by national policies on entry visas, but if the decisions were left to universities, I think most would take an open attitude with regard to students, while imposing a stringent filter on more senior scientific visitors.

One measure already being discussed would be an international agreement to make procurement or possession of dangerous pathogens, anywhere, an individual criminal offence in any country—just as hijacking of aircraft now is—and to foster a culture where "whistle blowing" is rewarded. A prime advocate of this campaign is Harvard professor Matthew Meselson, a leading expert on biological weapons.

Scientists are the critics of their subject, as well as the creators; quality control is enforced by the "peer review" that precedes publication of any new discovery in an academic journal. This is a safeguard against unmerited or exaggerated claims. But this copybook procedure is being violated more and more often, because of commercial pressures, or sometimes just intense academic rivalry. Newsworthy discoveries are trumpeted, via press releases or conferences, before they have been reviewed. In contrast, other discoveries are kept private, for commercial reasons. And scientists themselves face a dilemma when they are researching "sensitive" topics: lethal viruses, for instance.

One of the most spectacular departures from scientific

norms occurred in 1989 when Stanley Pons and Martin Fleischmann, then working at the University of Utah, claimed to have generated nuclear power at ordinary room temperature, using a tabletop apparatus. If credible, the claim fully merited all the hype it aroused: "cold fusion" would have offered the world an unlimited supply of cheap and clean energy. It would have ranked as one of the great discoveries of the century—indeed, one of the most momentous breakthroughs since the discovery of fire.

But technical doubts quickly surfaced. Extraordinary claims demand extraordinary evidence, and in this case the evidence proved far from robust. Inconsistencies were discerned in Pons and Fleischmann's claims; experimenters in several other laboratories tried to reproduce the phenomenon, but without success. Most scientists were sceptical and suspicious right from the start; within a year there was a general consensus that the results were misinterpreted, though even today there remain a few "believers."

A similar episode in 2002 was handled better. A group led by Rusi Taleyarkhan, a scientist at the Oak Ridge National Laboratory, was investigating a puzzling effect known as "sonoluminescence": when intense sound waves pass through a bubbly liquid, the bubbles get compressed and emit flashes of light. The Oak Ridge researchers claimed to have squeezed the imploding bubbles by a clever technique to such high temperatures that they became hot enough to trigger nuclear fusion, a fleetingly transient and miniaturised version of the process that keeps the Sun shining and generates the power in a hydrogen bomb. Not even all their colleagues at Oak Ridge believed them: the claim didn't violate "cherished beliefs" as much as cold fusion did, but still seemed implausible. But Taleyarkhan submitted a paper to the prestigious journal *Science*. Despite the scepticism of the referees, the editor chose to publish the paper but with an editorial warning that it was controversial. This decision at least ensured that the claim got maximal scrutiny.

The "cold fusion" fiasco did no great harm in the long run, except to the personal reputations of Pons and Fleischmann, and those who had jumped uncritically on their bandwagon. And the validity of Taleyarkhan's claims will soon be settled via debate and independent repetitions of his experiments. Any potentially epochal claim, provided it is openly announced, will be guaranteed to attract wide scrutiny from the international community of experts. So it doesn't matter a great deal if formal peer review is bypassed, provided that there is no impediment to openness.

Suppose, however, that a claim as extraordinary as Pons and Fleischmann's had come from senior scientists in a laboratory whose mission was military or commercially confidential research. What would then have happened? It is very unlikely that the work would ever have reached the public gaze: once the unprecedented economic and strategic importance of the "discovery" was appreciated by those in charge, a massive secret research programme would have got underway, consuming huge resources and shielded from open scrutiny.

Something very like this actually happened in the 1980s. The Livermore Laboratory, one of the two giant US laboratories involved in developing nuclear weapons, had a large secret programme aimed at producing x-ray lasers. This effort was funded as part of President Reagan's Strategic Defence Initiative ("star wars") project. The concept involved lasers in space that would be triggered by a nuclear explosion; in the microsecond before being vaporised the device was supposed to create intense "death rays" that could destroy incoming enemy missiles. Independent experts were almost uniformly scathing in their judgements. But it was the brainchild of Edward Teller and his protégés: working in a "closed" environment, with access to vast resources from the Pentagon, they were able to commit literally billions of dollars to this abortive "x-ray laser" scheme. Had one of Teller's scientists come up with a new source of energy, one can well imagine persuasive arguments, behind closed doors, that the national interest demanded a "crash" programme. In these examples, secrecy leads to waste, and to misdirection of effort. Even worse would be a clandestine project that actually posed risks that the experimenters were unaware of, or were downplaying, but which would have prompted most outside scientists to call for a halt.

"Fine-Grained Relinquishment"

An influential voice in favour of "going slow" is Bill Joy, co-founder of Sun Microsystems, and the inventor of the Java computer language. It was surprising to find such heartfelt unease expressed—in *Wired* magazine, of all places—by one of the heroes of cybertechnology, and his article "Why the Future Doesn't Need Us," published in 2000, attracted wide comment. The London *Times* ran an editorial likening it to the famous 1940 memorandum from the physicists Robert Frisch and Rudolf Peierls that alerted the UK government to the feasibility of an atomic bomb.

Joy's gaze is fixed on the far horizon. Rather than being fearful of where genetics and biotechnology might be leading us in the present decade—misapplications of genomics, the risk of bioterror by individuals, and so forth—Joy's disquiet focuses on the more remote threats of physics-based technologies. He is especially worried about the "runaway" consequences that may ensue when computers and robots surpass human capabilities. He isn't primarily concerned about malign misuse of the new technology, simply that genetics, nanotechnology, and robotics (GNR technologies) may develop uncontrollably and "take us over."

Joy's recipe is to "relinquish" the research and development that could make these threats real: "If we could agree, as a species, what we wanted, where we were headed, and why, then we would make our future much less dangerous—then we might understand what we can and should relinquish. Otherwise, we can easily imagine an arms race developing over GNR technologies, as it did with [nuclear] technologies in the twentieth century. This is perhaps the greatest risk, for once such a race begins, it's very hard to end it. This time—unlike during the Manhattan Project—we aren't in a war, facing an implacable enemy that is threatening our civilisation; we are driven, instead, by our habits, our desires, our economic system, and our competitive need to know."

As Joy realises, it wouldn't be easy to achieve a consensus that a specific type of research was so potentially dangerous that we should forgo it; human beings can seldom "agree, as a species"—the phrase Joy uses—even on what seem more urgent imperatives. Indeed, even a single enlightened individual would find it hard to know where to draw the line in research. So can relinquishment be sufficiently "fine grained" to discriminate between benign and hazardous projects? Novel techniques and discoveries will generally have manifest short-term usefulness, as well as being steps towards Joy's long-term nightmare. The same techniques that could lead to voracious "nanobots" might also be needed to create the nanotech analogue of vaccines that could immunise against them. If clandestine groups were pursuing dangerous research, it would be harder to devise countermeasures if nobody else had the relevant expertise.

Even if all the world's scientific academies agreed that some specific lines of inquiry had a disquieting "downside" and all countries, in unison, imposed a formal prohibition, then how effectively could it be enforced? An international moratorium could certainly slow down particular lines of research, even if they couldn't be stopped completely. When experiments are disallowed for ethical reasons, enforcement with ninety-nine percent effectiveness, or even just ninety percent, is far better than having no prohibition at all; but when experiments are exceedingly risky, enforcement would need to be close to one hundred percent effective to be reassuring: even one release of a lethal virus could be catastrophic, as could a nanotechnology disaster. Despite all the efforts of law enforcers, millions of people use illicit drugs; thousands peddle them. In view of the failure to control drug smuggling or homicides, it is unrealistic to expect that when the genie is out of the bottle, we can ever be fully secure against bioerror and bioterror: risks would still remain that could not be eliminated except by measures that are themselves unpalatable, such as intrusive universal surveillance.

My pessimism is nearer-term, and in some ways deeper, than Bill Joy's. He is concerned to stave off the day when superintelligent robots could take over from us, or the biosphere could crumble into "grey goo." But before these futuristic capabilities are attained, society could be dealt a shattering blow by misapplication of technology that exists already, or that we can confidently expect within the next twenty years. Ironically, the only compensating cheer is that if these shorter-term fears were realised, the hyperadvanced technology necessary for nanomachines and superhuman computers would suffer a perhaps irreversible setback, thereby safeguarding us against the scenarios that most trouble Bill Joy.

7

BASELINE NATURAL HAZARDS

Asteroid Impacts

We are at greater risk from a massive asteroid than from plane crashes, but the mounting human-induced threats are far more disquieting than any natural hazard.

IN JULY 1994, MILLIONS OF PEOPLE WATCHED, via the Internet, telescopic images of the largest and most dramatic "splashes" ever witnessed. Fragments of a large comet crashed into Jupiter; dark spots larger than the entire Earth, each a "scar" from a massive impact, were visible on that giant planet's surface for several weeks afterwards. The shattered comet, named Shoemaker–Levy after its discoverers, had been observed, in the previous year, to break up into about twenty pieces. Astronomers calculated that the fragments were on trajectories that would hit Jupiter, and geared up to watch the impacts at the predicted time.

This episode highlighted the vulnerability of our own planet to impacts of a similar kind. Earth presents a smaller target than Jupiter, the giant of our solar system, but comets and asteroids routinely come close enough to be a danger. About sixty-five million years ago Earth was hit by an object about ten kilometres across. The resultant impact released as much energy as a million H-bombs; it triggered mountain-shattering earthquakes around the world, and colossal tidal waves; it threw enough debris into the upper atmosphere to block out the Sun for more than a year. This is believed to have been the event that wiped out the dinosaurs. Earth still bears its scar: this momentous impact scoured out the Chicxulub crater in the Gulf of Mexico, nearly two hundred kilometres across.

Two separate classes of "rogue" objects hurtle around our solar system: comets and asteroids. Comets are made largely of ice, along with frozen gases such as ammonia and methane: they are often described as "dirty snowballs." Most comets spend nearly all their time, invisible to us, lurking in the cold outermost reaches of the solar system, far beyond even Neptune and Pluto; but sometimes they plunge inward towards the Sun on near-radial trajectories, getting warm enough for some ice to be vaporised, liberating gas and dust that reflects sunlight to create the conspicuous "tail." Asteroids, less volatile objects than comets, are composed of rocky material and move in near-circular orbits around the Sun. Most of them stay a safe distance from Earth, between the orbits of Mars and Jupiter. But some, the so-called near-Earth objects (NEOs), follow orbits that can intersect Earth's.

These NEOs range widely in size, from "minor planets" more than one hundred kilometres across, right down to mere pebbles. A ten-kilometre asteroid, harbinger of worldwide catastrophe and major extinctions, is expected to hit Earth once every only fifty to one-hundred-million years. The Chicxulub impact, sixty-five million years ago, may have been the most recent event of this magnitude. Two other similarly vast craters, one in Woodleigh, Australia, and another at Manicouagan, near Quebec, in Canada, could be the aftermaths of comparable impacts 200 to 250 million years ago. Perhaps one of these caused the greatest extinctions of all, at the Permian/Triassic transition 250 million years ago. (At the time of these impacts, the Atlantic Ocean had not opened up, and most of Earth's land mass was part of a single continent, known as Pangaea.)

Smaller asteroids (and less devastating impacts) are much more common: NEOs one kilometre across are a hundred times more numerous than the extinction-triggering ten-kilometre asteroids; one-hundred-metre-sized bodies are probably a hundred times more numerous still. The well-known Barringer crater in Arizona was carved out by an asteroid about a hundred metres across, which hit about fifty thousand years ago; a similar crater in Wolfe Creek, Australia, is about three hundred thousand years old. NEOs fifty metres across seem to hit Earth once per century. In 1908, the Tunguska meteorite devastated a remote part of Siberia. It was moving so fast, up to forty kilometres per second, that its impact packed the punch of a forty-megaton explosion. It vaporised and exploded high in the atmosphere, flattening thousands of square kilometres of forest but leaving no crater.

A Low Risk, but Not Negligible

We do not know whether a large dangerous NEO "with our name on it" is destined to hit in the coming century. However, we know enough about how many asteroids there are on

Earth-crossing orbits to be able to quantify the probability. The risk isn't large enough to keep anyone awake at night, but it isn't completely negligible either. There is a fifty percent risk of a Tunguska-scale impact somewhere on Earth this century. But most of Earth's surface is either covered with oceans or sparsely populated, so the chance of an impact on a densely populated region is far smaller: but such an event could cause millions of fatalities.

In the world as a whole, the risk from floods, tornados and earthquakes looms larger. (Indeed the worst localised natural catastrophe that could be deemed probable in this century would be an earthquake in Tokyo or perhaps in Los Angeles, where the immediate devastation would have a longer-term "fallout" for the world's economy.) However, for Europeans and North Americans outside the areas most prone to earthquakes or hurricanes, asteroid impact is actually the number-one natural hazard. The dominant risk is not from Tunguska-scale events, but from rarer impacts that would each devastate a larger area.

If you are now, say, twenty-five years old, your future life expectancy is about fifty years. The chance of being a victim of a massive asteroid impact is therefore roughly the probability that one happens in the next fifty years. Before that time is up, there is about one chance in ten thousand that an asteroid half a kilometre across will crash in to the North Atlantic, causing giant tsunamis (tidal waves) that would destroy the North American and European seaboard; or into the Pacific, where it would have similar consequences for the coasts of East Asia and the Western US. The probability that we'll end our lives (along with many millions of others) in such an event is about the same as the average person's risk of dying in an air crash—somewhat higher, indeed, if we live near a coast, where we are vulnerable to smaller tsunami.

It is a minor risk, but no lower than other hazards that governments take measures to guard against or ameliorate. A recent report on NEOs commissioned by the British government presented the situation like this: "If a quarter of the world's population were at risk from the impact of an object of 1 kilometre diameter, then according to current safety standards in use in the United Kingdom, the risk of such casualties, even if occurring on average once every 100,000 years, would significantly exceed a tolerable level. If such risks were the responsibility of an operator of an industrial plant or other activity, then that operator would be required to take steps to reduce the risk."

By discovering and tracking the most dangerous Earth-crossing NEOs, we could in principle have years of warning of any major catastrophe. Were an impact in mid-Atlantic to be forecast, then a massive evacuation of coastal areas could save tens of millions of lives, even if we couldn't do anything to divert the incoming object. The international community spends billions of dollars a year on weather forecasting, and can thereby predict hurricanes; it would seem worth a few millions to ensure that a (much more unlikely but far more devastating) giant tsunami—as portrayed in the movie *Deep Impact*—didn't catch us unawares.

Reducing the Risk?

There is another motive for surveying and cataloguing all NEOs: in the long run, it may be possible to deflect rogue objects away from Earth, but very accurately known orbits are a prerequisite, and accuracy will not be achieved unless these objects have been tracked for a long time beforehand. Arthur C. Clarke's novel *Rendezvous with Rama* describes how a Tunguska-type event wipes out northern Italy. (The year Clarke picked for this catastrophe was 2077, and the date, coincidentally, September 11.) "After the initial shock, mankind reacted with a determination and a unity that no earlier age could have shown. Such a disaster might not occur again for a thousand years—but it might occur tomorrow. Very well; there would be no next time. No meteorite large enough to cause catastrophe would ever again be allowed to breach the defences of Earth. So began Project Spaceguard."

"Spaceguard"-type projects, whereby we can not merely be forewarned but even protected against asteroid impacts, need not remain science fiction: they could be implemented within fifty years. If we knew several years in advance that an NEO was on course to hit Earth, nothing could be done about it today. But within a few decades we might have the technology needed to divert the trajectory enough to ensure that the "rogue" object posed no danger. The longer the advance warning we had of an impending impact, the smaller would be the orbital nudge needed to change its course so that it missed us. But it would be imprudent even to attempt such an enterprise without knowing a great deal more about what asteroids are made of than we now do. Some are solid boulders; but others (perhaps most) may be loosely packed piles of rock held together only by "stickiness" and by their very weak gravity. In the latter case, attempts to push an asteroid off course (especially by drastic methods such as a nuclear explosion) could shatter it into pieces, which would pose an even greater aggregate risk to Earth than the original single body did.

Comets are harder to deal with. A few (like Halley's comet) return repeatedly and follow well-charted orbits, but most approach us "cold" from deep space, giving no more than a year's warning. Also, their orbits are somewhat erratic because gas squirts from them, and fragments break off in unpredictable ways. For these reasons, comets pose an intractable and perhaps irreducible risk to us.

A numerical index of the seriousness of unlikely catastrophes such as potential asteroid impacts was introduced by Richard Binzel, a professor at MIT. This was adopted at an international conference in Turin (Torino) and has become known as the Torino scale. It resembles the familiar Richter scale for earthquakes. However, an event's ranking on the Torino scale takes account of its likelihood as well as its magnitude: the seriousness of a potential threat depends on its probability, multiplied by the amount of devastation that would ensue if it actually happened. The scale runs from 1 to 10. A fifty-metre asteroid, like the one that exploded above Siberia in 1908, would have rank 8 on the scale if it were certain to hit us; a one-kilometre asteroid would have rank 10 if it were certain to hit, but only 8 if its orbit was so poorly known that we could merely predict that it would pass somewhere within a million kilometres of Earth. Earth is only 12,750 kilometres across, so the probability of hitting the "bull's-eye" would then be about one in ten thousand.

The Torino number assigned to a particular event can change as our evidence mounts. For example, the path of a hurricane may initially be hard to predict; as it progresses, we can forecast with ever-increasing confidence whether it will pass over a populous island or whether it will miss. Likewise, the longer we track an NEO, the more precisely we can predict its future trajectory. Large asteroids are regularly identified that on the basis of a crudely determined orbit could endanger Earth. But when their orbits have been pinned down more exactly, we generally become confident that they will miss, so their ranking on the Torino scale drops towards zero. However, in the minority of cases where the area of uncertainty shrinks, but Earth remains within it, we would have reason to become even more worried, and the Torino number would rise, perhaps from 8 towards 10.

Experts in NEO impacts have now devised a more refined index, called the Palermo scale, which takes account of how far in the future the possible impact would occur. This is a better measure of how concerned we should be. For instance, if we knew that a fifty-metre asteroid would hit Earth next year, that would rate high on the Palermo index, but if the impact of that particular object were forecast, with an equally high confidence level, for (say) the year 2890, it would not raise our anxiety level at all. This is not simply because we discount future risks (especially if they are so far ahead that we are all dead) but because the law of averages leads us to expect several Tunguska-scale events, caused by similarly sized asteroids, before that time.

Modest efforts are worthwhile to monitor the few thousand largest NEOs that could pose a threat. If the conclusion was that none of them would hit Earth in the next fifty years, we would have gained a degree of reassurance that would be worth the modest collective investment involved. If the outcome were less reassuring, we could at least prepare ourselves; moreover, if the predicted impact were (say) fifty years from now, there might be enough time to develop the technology to deflect the rogue object. It is also worth improving statistical knowledge of the smaller objects, even though we could not expect to have much forewarning if one of these were heading towards a direct hit with Earth.

Supereruptions

Apart from the ever-present hazard from impacting asteroids and comets there are other natural catastrophes even harder to predict far ahead, and that are even more difficult to prevent or stave off: extremely violent earthquakes and volcanic eruptions, for instance. The latter include a rare class of "supereruptions," thousands of times larger than the eruption of Krakatoa in 1883, that would propel thousands of cubic kilometres of debris into the upper atmosphere. A crater in Wyoming, eighty kilometres across, is a relic of such an event about a million years ago. Rather closer to the present, a supereruption in northern Sumatra seventy thousand years ago left a one-hundred-kilometre crater and ejected several thousand cubic kilometres of ash, enough to have blocked out the Sun for a year or more.

Two aspects of these violent natural catastrophes are, however, somewhat reassuring. Firstly, massive asteroid impacts and colossal volcanic eruptions are so rare that reasonable people aren't deeply anxious about them, nor preoccupied with them (though, were it technically feasible, it would be worth a substantial investment to reduce the risk further). Secondly, they are not getting any worse: we may be more aware of them than earlier generations were (and society is certainly more risk-averse than it was), but nothing humankind does is likely to increase the risk of asteroid impacts, nor of volcanic super-eruptions.

They serve therefore as a "calibration" against the fast-growing human-induced risks to the environment, which could according to pessimistic scenarios become thousands of times larger.

8

HUMAN THREATS TO EARTH

---

*Environmental changes induced by human
activities, still poorly understood, may be
graver than the "baseline" threats from
earthquakes, eruptions, and asteroid impacts.*

IN HIS BOOK *THE FUTURE OF LIFE,* E.O. Wilson sets the
scene with an image that highlights the complex fragility of
"Spaceship Earth": "The totality of life, known as the biosphere to scientists and creation to the theologians, is a membrane of organisms wrapped around Earth so thin it cannot be
seen edgewise from a space shuttle, yet so internally complex
that most species composing it remain undiscovered."

Human beings are depleting the variety of Earth's plant and
animal life. Extinctions are, of course, intrinsic to evolution
and natural selection: fewer than ten percent of all the species
that ever swam, crawled, or flew are still on Earth today. An

extraordinary procession of species (almost all now extinct) has traced the higgledy-piggledy path by which natural selection led from unicellular organisms to our present biosphere. For over a billion years, primitive "bugs" exhaled oxygen, transforming the young Earth's poisonous (to us) atmosphere and clearing the way for complex multicellular forms of life—relative newcomers—and for our eventual emergence.

It needs an imaginative leap to grasp geological time spans, and how colossally prolonged they are compared to hominoid history, which is in turn far longer than recorded human history. (In popular culture these huge disjunctions are sometimes elided, as in old movies like *A Million Years BC*, portraying Raquel Welch cavorting among the dinosaurs.)

We know from fossils that a cornucopia of swimming and creeping things evolved during the Cambrian era 550 million years ago leading to a vast diversification of species. The next 200 million years saw the greening of the land, offering a habitat for exotic creatures: dragonflies as big as seagulls, millipedes a yard long, giant scorpions and squid-like sea monsters. Then came the dinosaurs. Their sudden demise sixty-five million years ago opened the way to mammals, to the emergence of apes and ourselves. A species lasts for millions of years; even the most rapid bursts of natural selection generally take thousands of generations to change the appearance of any species. (Catastrophic events can, of course, induce drastic changes in animal populations; asteroid impacts, for instance, can trigger sudden extinctions.)

The Sixth Extinction

Geological records reveal five great extinctions. The largest of all happened at the Permian/Triassic transition around 250

million years ago; the second largest, 65 million years ago,
wiped out the dinosaurs. But human beings are perpetrating a
"sixth extinction" on the same scale as earlier episodes. Species
are now dying out at one hundred or even one thousand times
the normal rate. Before *Homo sapiens* came on the scene, about
one species in a million became extinct each year; the rate is
now is closer to one species in a thousand. Some species are being killed off directly; but most extinctions are an unintended
outcome of human-induced changes in habitat, or of the introduction of nonindigenous species into an ecosystem.

Biodiversity is being eroded. Extinctions are deplorable not
just for aesthetic and sentimental reasons, attitudes engendered
disproportionately by the so-called charismatic vertebrates—
the tiny minority of species that are feathered, furry, or grandly
oceanic. Even at the most utilitarian level, we are destroying
the genetic variety that may prove of value to us. As Robert
May says, "We are burning the books before we've learnt to
read them." Most species have not yet even been catalogued.
Gregory Benford has proposed a Library of Life project, an urgent effort to gather, freeze, and store a sample of the complete
fauna of a tropical rain forest, not as a substitute for conservation measures, but as an "insurance policy."

Threats to the biosphere are becoming ever greater with
biotechnical advances. For instance, salmon on fish farms, genetically modified to grow faster and larger, could outcompete
natural varieties if they escaped into the wild. Worst of all, new
diseases, unwittingly released, could devastate species. Above
all, this impending diminution of nature's riches connotes a
failure of our stewardship of the planet.

But the yearning for an unspoilt "natural" world is naive. The
environment many of us cherish and feel most attuned to—in
my case the English countryside—is an artificial creation, the
outcome of centuries of intensive cultivation, enriched by many

nonindigenous plants and trees introduced by farmers and gardeners. Even the North American landscape of the "Old West" is far from natural. The Indians had been transforming the terrain long before the first invasions from Europe: "slash and burn" had been practised for at least a millennium, rendering the country far more open and less forested than its pristine state. The land was even more intensively transformed in the twentieth century.

Population Projections

Humanity's long-term impact on Earth depends both on population and on lifestyle. WWF, a conservation group, has published estimates of the land area, or "footprint," needed to support each person: it concludes that an area equivalent to "almost three planets" would be required to support the world population with the lifestyle and consumption pattern that it predicts for 2050. This particular calculation is controversial and perhaps somewhat tendentious: for instance, the "footprint" includes the area of forest needed to soak up the carbon dioxide arising from each person's energy use, making no allowance for a shift to renewable energy sources, nor for the tenable viewpoint that modest rises in carbon dioxide levels are tolerable. Nonetheless, the world plainly could not perpetually support its entire population in the present style of middle-class Europeans and North Americans.

At the other extreme, a population as high as ten billion would be fully sustainable if everyone lived in tiny apartments, perhaps like the "capsule hotels" that already exist in Tokyo, subsisting on a rice-based vegetarian diet, electronically networked, travelling little, and finding recreation and fulfilment in virtual reality rather than the consumerism and incessant travel now favoured in the profligate West. Such a lifestyle would be frugal in its demands on energy and natural resources. It need not, however, be incompatible with cultural and technical advance: indeed, the most dramatic engines of current economic growth—miniaturisation and information technology—are environmentally benign.

For a population to remain in a steady state, women must each have, on average, 2.1 children (the extra 0.1 accounting for children who never reach reproductive age). The fertility rates in many developed countries are well below this. Perhaps surprisingly, Catholic Italy has the lowest rate of all—only 1.2 births per woman. Almost equally low are Greece and Spain, along with Russia and Armenia.

This drastic reduction in family size is not just a European phenomenon. There are now more than sixty countries where fertility is below replacement level. These include not only China, where there has long been insistent political pressure for "one-child families," but other Asian countries such as Japan, Korea, and Thailand, were there has not been such pressure. And there have been drastic declines elsewhere. For instance, despite the anticontraception policy of the Catholic Church, the fertility rate in Brazil has halved in twenty years, and is now 2.3. In Iran, where the ruling mullahs in the 1990s were openly hostile to the UN's agenda for limiting population growth, women have made their own decisive choices, and the fertility rate has fallen from 5.5 in 1988 to 2.2 today.

Despite the low birth rate, the population of Europe is still rising, partly because the children of the "population explosion" are now of childbearing age, and also because of immigration and improved life expectancy. Medical advances and public health measures have extended life expectancy and robustness in all but the most deprived parts of the world.

Without an intervening catastrophe, world population still

seems destined to continue rising until 2050, by which time it will have reached eight billion. This projection follows from the fact that the present age distribution in developing countries is sharply skewed towards the young, so an increase would continue even if these people had less than the replacement level of children. This increase, combined with the trend towards urbanisation, will lead to at least twenty "megacities" with populations exceeding twenty million.

But the surprisingly rapid fall in fertility, stemming from the empowerment of women, has led the UN to reduce its projections for the second half of this century. The best current guess is that after 2050 the population will start to drop, perhaps falling back to its present-day value by the end of the century, unless medical advances boost life expectancy to the extent that some futurologists predict. The "over-fifties" will dominate in Europe and North America, even without any novel techniques for extending life span. This trend may be masked, particularly in the US, by immigration from the developing world, where stabilisation and consequent decline (if it ever happens) will be delayed.

Of course, the extrapolation is based on assumptions about social trends. If European countries became genuinely anxious about falling population, governments could readily introduce measures to stimulate fertility. Contrariwise, epidemics spreading within megacities could cause catastrophic declines of the kind already projected for parts of Africa, and by 2050 such predictions could be radically changed by technical advances in robotics and medicine as drastic as those that techno-enthusiasts envisage.

The most benign outcome, if we could indeed survive the next century without catastrophic reversals, would be a world with a population lower than at present (and far below its projected peak in around 2050).

A new hazard that must be folded into these projections, and perhaps a portent of others, is the AIDS epidemic. This didn't catch hold in the human population until the 1980s, and it still hasn't reached its peak. Almost ten percent of South Africa's forty-two million people are believed to be HIV positive: AIDS is predicted to cause seven million deaths by 2010 in that country alone, wiping out much of the most productive age group, cutting the life expectancy of both men and women by as much as twenty years, and leaving millions of traumatised orphans among the younger generation. The burgeoning AIDS pandemic will devastate Africa; millions of cases are projected in Russia; the total numbers infected are rising fast in China and India, where fatalities from AIDS may exceed African levels within a decade.

Can we expect other calamitous "natural" plagues? Some experts have been reassuring about our likely susceptibility. Paul W. Ewald, for instance, notes that global migrations, and the consequent mixing of people over the last century, have exposed everyone to pathogens from all parts of the world, but there has been only one devastating pandemic: HIV-AIDS. The other naturally occurring viruses, like ebola, are not durable enough to generate a runaway epidemic. But Ewald's mildly upbeat appraisal leaves aside the risk of some epidemic triggered by bioerror or bioterror, rather than by nature.

Earth's Inconstant Climate

Climatic change has, like extinction of species, characterised Earth throughout its history. But it has, like the extinction rate, been disquietingly speeded up by human actions.

The climate has undergone natural changes on every time scale, from decades to hundreds of millions of years. Even within the era of recorded history the regional climate has varied markedly. It was warmer in Northern Europe a thousand years ago: there were agricultural settlements in Greenland where animals grazed on land that is now ice-covered; and vineyards flourished in England. But there have been prolonged cold periods too. The warm spell seems to have ended by the fifteenth century, to be succeeded by a "little ice age" that continued until the end of the eighteenth century. There are regular records of the ice on the Thames getting so thick during much of that period that fires were lit on it; glaciers in the Alps advanced. The "little ice age" may offer important clues to a question that has been perennially controversial: whether variability of the Sun could trigger alterations in the climate. During this cold spell, the Sun seemed to have been behaving slightly erratically: in the second half of the seventeenth century and the first years of the eighteenth there was a mysterious seventy-year period, (now known as the Maunder minimum, after the scientist who first noticed it) in which there were hardly any sunspots at all. The activity on the Sun's turbulent surface—flares, sunspots and so forth—normally rises to a peak and then drops again, repeating this cycle somewhat unsteadily, but roughly every eleven or twelve years. Claims that this cycle affects the climate date back more than two hundred years, but are still controversial. (It has even been alleged that the economic cycle "tracks" solar activity.) There are also claims that the length of a particular cycle—whether it is closer to eleven years or twelve years—affects the average temperature.

Nobody really understands how sunspots and flaring activity (or their absence) could affect the climate to this extent. Sunspots are linked to the magnetic behaviour of the Sun, and to the flares that generate fast-moving particles that hit Earth. These particles themselves, however, carry only a tiny fraction of the Sun's energy, but we should be open-minded about the possibility that some "amplifier" in the upper atmosphere might enable them to trigger substantial changes in cloud cover. Scientists have often been caught out in the past, rejecting evidence staring them in the face because they couldn't at the time think of how to explain it. (A spectacular instance of this is continental drift. The coastline of Europe and Africa seems to fit that of the Americas, like a jigsaw puzzle, as though these landmasses had once been joined and had drifted apart. Until the 1960s, nobody understood how the continents could move, and some leading geophysicists denied the evidence of their own eyes rather than accept that continental motion might be induced by some mechanism that they hadn't been astute enough to think of.)

There are other environmental effects on climate, such as major volcanic eruptions. The 1815 eruption of the Tambora volcano in Indonesia propelled about one hundred cubic kilometres of dust into the stratosphere, along with gases that combined with water vapour to create an aerosol of sulphuric acid droplets. Exceptionally cold weather the following year, both in Europe and in New England, led to 1816 being termed "the year without a summer." (Mary Shelley wrote her gothic fantasy *Frankenstein*—the first modern science fiction novel—during that year's unseasonable weather, while holed up in Byron's rented villa on the shores of Lake Geneva.)

One human-induced atmospheric change that was completely unpredicted was the emergence of the ozone hole over the Antarctic caused by chemical reactions of chlorofluorocarbons (CFCs) in the stratosphere that depleted the ozone layer. International agreement to phase out the culprit CFCs, used in aerosol cans and as a coolant in domestic refrigerators, has ameliorated this problem: the ozone hole is now filling in. But we were actually lucky that this problem was so readily remedied. Paul Crutzen, one of the chemists who elucidated how

CFCs actually acted in the upper atmosphere, has pointed out that it was a technological accident and quirk of chemistry that the commercial coolant adopted in the 1930s was based on chlorine. Had bromine been used instead, the atmospheric effects would have been more drastic and longer-lasting.

Greenhouse warming

In contrast to ozone depletion, global warming due to the so-called "greenhouse effect" is an environmental problem for which there is no quick fix. This effect comes about because the atmosphere is more transparent to incoming sunlight than to the infrared "heat radiation" that is emitted by the Earth; the heat is therefore trapped, rather as in a greenhouse. Carbon dioxide is one of the "greenhouse gases" (water vapour and methane being others) that trap the heat. Atmospheric carbon dioxide is already fifty percent above its pre-industrial level, because of the increasing consumption of fossil fuels. There is a consensus that this accumulation will make the world hotter in the twenty-first century than it would otherwise be, but exactly how much hotter is still unclear. The main temperature rise is likely to be between two and five degrees. Few venture more precise predictions; many warn that even more extreme scenarios cannot be ruled out. Even if the rise were just two degrees, a very conservative estimate, there could be serious localised consequences (e.g., more storms and other extreme weather).

There is nothing optimal about the Earth's present climate: it is simply something to which human civilization has accommodated over the centuries, as have the animals and plants (both natural and agricultural) with which we share the terrain. The reason that the impending global warming could be threateningly disruptive is that it will happen much more rapidly than the naturally occurring changes in the historical past; too fast for human populations, land-use patterns, and natural vegetation to adjust to. Global warming may induce a rise in sea level, an increase in severe weather, and a spread of mosquito-borne diseases to higher latitudes. On the bright side (from our human perspective) the climate in Canada and Siberia will become more temperate.

Steady global warming at the "conservative best guess" rate would impose costs in agricultural adjustments, sea defenses and other areas, and aggravate droughts in some regions. Concerted action by governments to reduce global warming is certainly worthwhile. It would be an exaggeration, however, to regard a temperature rise of two or three degrees as in itself a global catastrophe. It would be a setback to economic advance, and an impoverishment of many nations. Famines within a country most often arise from maldistribution of wealth, rather than an overall shortage of food, and can be ameliorated by government action. Likewise, the consequences of climate change could be softened, and distributed more equitably, by international action.

The apparent slow-down in population growth is of course good news for global warming scenarios: fewer people mean less emission. But there is so much inertia in the atmospheric and ocean systems that, whatever happens, a rise of at least two degrees in mean temperature by 2100 seems likely. Any projections beyond that time obviously depend on how large the population would be, and how people live and work. Even more, the long-range prognosis will depend on whether fossil fuels are replaced by alternative sources of energy. Optimists hope that this will happen as a matter of course. The anti-gloom environmental propagandist Bjorn Lomberg quotes a Saudi-Arabian oil minister's dictum that "the oil age will end, but not for lack of oil, just as the stone age ended, but not for lack of stones." But most experts believe government-imposed ceilings on carbon dioxide emission are worthwhile not only for their direct impact, but as a stimulus to development of more efficient renewable energy sources.

What Are the "Worst Cases?"

For the bulk of the world's population, the twentieth century ideological stances of East–West relations that motivated the nuclear confrontation were an irrelevant distraction from the immediate problems of poverty and environmental hazard. To the age-old "threats without enemies" (earthquake, storm and drought) must now be added the man-made threats to the biosphere and the oceans. The Earth's biosphere has of course changed ceaselessly over its history. But the current changes—pollution, loss of biodiversity, global warming, etc.—are unprecedented in their speed.

The problems of environmental degradation will become far more threatening than they even are today. The ecosystem may not be able to adjust to them. Even if global warming occurs at the slower end of the likely range, its consequences—competition for water supplies, for example, and large-scale migrations—could engender tensions that trigger international and regional conflicts, especially if these are further fuelled by continuing population growth. Moreover, such conflict could be aggravated, perhaps catastrophically, by the increasingly effective disruptive techniques with which novel technology is empowering even small groups.

The interaction of atmosphere and oceans is so complex and uncertain that we can't discount the risk of something much more drastic than the "best guess" rate of global warming. The rise by 2100 could even exceed five degrees. Even worse, the temperature change may not be just in direct (or "linear") proportion to the rise in the carbon dioxide concentration. When some threshold level is reached, there could be a sudden and drastic "flip" to a new pattern of wind and ocean circulation.

The Gulf Stream is part of a so-called "conveyor-belt" flow pattern whereby warm water flows north-eastward towards Europe near the surface, and returns, having cooled, at greater depths. The melting of Greenland's ice would release a huge volume of fresh water, which would mix with the salt water, diluting it and rendering it so buoyant that it would not sink even after it had cooled. This injection of fresh water could thereby quench the "thermohaline" circulation pattern (controlled by the ocean's salinity and temperature) that is crucial for maintaining the temperate climate of northern Europe. If the Gulf Stream were truncated or reversed, Britain and neighboring countries could be plunged into near-arctic winters, like those that currently prevail in similar latitudes in Canada and Siberia.

We know that changes of this kind happened in the past because cores drilled through the ice sheets of Greenland and the Antarctic provide a kind of fossil record of temperatures: each year fresh ice freezes on top and squeezes down the earlier layers. Many times during the last hundred thousand years there seem to have been drastic cooling-offs within decades or less. The climate has actually been unusually stable during the last eight thousand years. The worry is that human-induced global warming may render the next "flip" much more imminent.

"Flipping" of the Gulf Stream would be a disaster for Western Europe, even though it might have a countervailing "upside" elsewhere. Another scenario (admittedly unlikely) would be a so called "runaway greenhouse effect" where rising temperatures cause a positive feedback that releases still more greenhouse gas. Earth would need to be already substantially hotter than it actually is to be at any risk from runaway evaporation of water from the oceans (water vapour being a greenhouse gas). But we cannot so firmly exclude a runaway due to the release of huge amounts of methane (at least twenty times as efficient as carbon dioxide as a greenhouse gas) trapped in the soil. Such a runaway would be a global disaster.

If we could be absolutely sure that nothing more drastic than "linear" changes in the climate could occur, it would be reassuring. The small chance of something really catastrophic is more worrying than the greater chance of less extreme events. Not even the most drastic conceivable climatic shifts could directly destroy all humanity, but the worst of them, accompanied by transitions to far more variable and extreme weather patterns, could negate decades of economic and social advance.

Even a one-percent chance that human-induced atmospheric changes could trigger an extreme and sudden climatic transition—and a meteorologist would need to be very confident indeed to set the odds as low as that—is a disquieting enough prospect to justify precautionary measures more drastic than those already proposed by the Kyoto agreements (which require industrialized countries to reduce their carbon dioxide emissions to 1990 levels by 2012). Such a threat would be a hundred times larger than the baseline risk of environmental catastrophe that the Earth is exposed to, irrespective of human actions, from asteroid impacts and extreme volcanic events.

I conclude this chapter with a sober assessment from Charles, Prince of Wales, whose views are seldom quoted approvingly by scientists: "The strategic threats posed by global environment and development problems are the most complex, interwoven and potentially devastating of all the challenges to our security. Scientists . . . do not fully understand the consequences of our many-faceted assault on the interwoven fabric of atmosphere, water, land and life in all its biological diversity. Things could turn out to be worse than the current scientific best guess. In military affairs, policy has long been based on the dictum that we should be prepared for the worst case. Why should it be so different when the security is that of the planet and our long-term future?"

9

EXTREME RISKS
A Pascalian Wager

Some experiments could conceivably threaten the entire Earth. How close to zero should the claimed risk be before such experiments are sanctioned?

THE MATHEMATICIAN AND MYSTIC Blaise Pascal offered a famous argument for devout behaviour: even if you thought it exceedingly unlikely that a vengeful God existed, you would be prudent and rational to behave as though He did, because it is worth paying the (finite) price of forgoing illicit pleasures in this life as an "insurance premium" to guard against even the smallest probability of something infinitely horrible—eternal Hellfire—in the afterlife. This argument seems to carry little resonance today, even among proclaimed believers.

Pascal's celebrated "wager" is an extreme version of the "precautionary principle." This line of reasoning is widely invoked in health and environmental policy. For example, the long-term consequences of genetically modified plants and animals for human health, and for ecological balance, are manifestly uncertain: a calamitous outcome may seem improbable, but we cannot say that it is impossible. Proponents of the precautionary principle urge that we should proceed cautiously, and that the onus should be on the advocates of genetic modification to convince the rest of us that such fears are ungrounded—or, at the very least, that the risks are small enough to be outweighed by some specific and substantial benefits. An analogous argument is that we should forgo the benefits of extravagant energy consumption, and thereby reduce the deleterious consequences of global warming—especially the small risk that its consequences could be far more serious than the "best guess" suggests.

The obverse of technology's immense prospects is an escalating variety of potential disasters, not just from malevolent intent but from innocent inadvertence as well. We can conceive of events—albeit unlikely ones—that could cause worldwide epidemics of fatal diseases to which there is no antidote, or change society irreversibly. And robotics and nanotechnology could in the long term be even more threatening.

However, it is not inconceivable that physics could be dangerous too. Some experiments are designed to generate conditions more extreme than ever occur naturally. Nobody then knows exactly what will happen. Indeed, there would be no point in doing any experiments if their outcomes could be fully predicted in advance. Some theorists have conjectured that certain types of experiment could conceivably unleash a runaway process that destroyed not just us but Earth itself. Such an event seems far less likely than the human-induced bio- or nanocatastrophes that could befall us during this century—less likely, indeed, than a massive asteroid impact. But if such a disaster occurred, it would by any reckoning be worse than "merely" destroying civilisation, or even destroying all human life. It raises the issue of how we quantify relative degrees of awfulness, and what precautions should be taken (by whom) against occurrences that might seem to have an infinitesimal probability, but which could lead to an "almost infinitely bad" calamity. Should we forgo some kinds of experiments, for the same reason that Pascal recommended prudent behaviour?

Risking the Earth

Promethean concerns of this kind go back to the atomic bomb project during the Second World War. Could we be absolutely sure, some then wondered, that a nuclear explosion wouldn't ignite all the world's atmosphere or oceans? Edward Teller contemplated this scenario as early as 1942, and Hans Bethe made a quick calculation that seemed reassuring. Before the 1945 "Trinity" test of the first atomic bomb in New Mexico, Teller and two colleagues addressed the question in a Los Alamos report. The authors focused on a possible runaway reaction of atmospheric nitrogen, and wrote that "the only disquieting feature is that the 'safety factor' decreases rapidly with initial temperature." This inference led to renewed concern in the 1950s, because hydrogen (fusion) bombs indeed generate even higher temperatures; another physicist, Gregory Briet, revisited the problem before the first H-bomb test. It is now clear that the actual "safety factor" was very large indeed. One nonetheless wonders how small the contemporary estimates of that factor would have needed to be before those in charge would have felt it prudent to abandon the H-bomb tests.

We now know for certain that a single nuclear weapon, devastating though it is, cannot trigger a nuclear chain reaction that would utterly destroy Earth or its atmosphere. (The entire arsenals of the US and the Russia, were they to be unleashed, could nevertheless have an effect as bad as any natural disaster that could be expected in the next hundred thousand years.) But some physics experiments carried out for reasons of pure scientific inquiry could conceivably—or so some have claimed—pose a global, even cosmic, threat. These experiments offer an interesting "case study" of who should decide (and how) whether to sanction an experiment with a catastrophic "downside" that is very improbable but not quite inconceivable, especially when the leading experts may not have enough confidence in their theories to offer the compelling level of reassurance that the public might properly expect.

Most physicists (and I would count myself among them) would rate these threats as very, very improbable. But it is important to make clear what such a rating actually means. There are two distinct meanings of probability. The first, leading to a firm and objective estimate, applies when the underlying mechanism is well understood, or when the event being studied has happened many times in the past. For instance, it is easy to work out that when an unbiased coin is tossed ten times the chance of getting ten heads is a bit less than one in a thousand; and the chance of catching measles during an epidemic may also be quantified, because even though we may not understand all the biological details of virus transmission, we have data on many earlier epidemics. But there is a second kind of probability that reflects no more than an informed guess, and may alter as we learn more. (The assessments that different experts give of, for instance, the consequences of global warming are "subjective likelihood" estimates of a similar kind.) In a criminal investigation, the police may say that it "seems very probable" or "is highly improbable" that a body is buried in a particular place. But this reflects just the betting odds they would offer in the light of the available evidence. Further digging will reveal that the body either is or isn't there, and the probability is thereafter either one or zero. When physicists contemplate an event that has never happened before, or a process that is poorly understood, any assessment they can offer resemble this second kind of probability: it is an informed guess, buttressed (often very strongly) by well-established theories but nonetheless open to revision in the light of new evidence or insight.

Our "Final" Experiment?

Physicists aim to understand the particles that the world is made of, and the forces that govern those particles. They are eager to probe the most extreme energies, pressures, and temperatures; for this purpose they build huge and elaborate machines: particle accelerators. The optimum way to produce an intense concentration of energy is to accelerate atoms to enormous speeds, close to that of light, and crash them together. It is best of all to use very heavy atoms. A gold atom, for instance, has nearly two hundred times the mass of a hydrogen atom. Its nucleus contains 79 protons and 118 neutrons. A lead nucleus is heavier still, containing 82 protons and 125 neutrons. When two such atoms are crashed together, their constituent protons and neutrons implode to a density and pressure far higher than what they were when they were packed into a normal gold or lead nucleus. They may then break up into still smaller particles. According to theory, each proton and neutron consists of three quarks, so the resultant "splat" releases over a thousand quarks. These ultra-fast atomic collisions actually replicate, in microcosm, those that prevailed in the first microsecond after the "big bang," when all the matter in the universe was squeezed into a so-called quark–gluon plasma.

Some physicists raise the possibility that these experiments might do something far worse than smashing a few atoms, like destroying our Earth, or even our entire universe. Such an event is the theme of Greg Benford's novel *COSM*, where an experiment at Brookhaven laboratory devastates the accelerator and creates a new "microuniverse" (which remains, comfortingly, encased within a sphere small enough to be carried around by its graduate-student creator).

An experiment that generates an unprecedented concentration of energy could—conceivably, but highly implausibly—trigger three quite different disaster scenarios.

Perhaps a black hole could form, and then suck in everything around it. According to Einstein's theory of relativity, the energy needed to make even the smallest black hole would far exceed what these collisions could generate. Some new theories, however, invoke extra spatial dimensions beyond our usual three; a consequence would be to strengthen gravity's grip, rendering it less difficult than we previously thought for a small object to implode into a black hole. But the same theories suggest that these holes would still be innocuous, because they would erode away almost instantly, rather than tugging in more stuff from their surroundings.

The second frightening possibility is that the quarks might reassemble themselves into a very compressed object called a strangelet. That in itself would be harmless: the strangelet would still be far smaller than a single atom. However, the danger is that a strangelet could, by contagion, convert anything else it encountered into a strange new form of matter. In Kurt Vonnegut's novel *Cat's Cradle* a Pentagon scientist produces a new form of ice, "ice nine," that is solid at ordinary temperatures; when it escapes from the laboratory it "infects" natural water, and even the oceans solidify. Likewise, a hypothetical strangelet disaster could transform the entire planet Earth into an inert hyperdense sphere about one hundred metres across.

The third risk from these collision experiments is still more exotic, and potentially the most disastrous of all: a catastrophe that engulfs space itself. Empty space—what physicists call "the vacuum"—is more than just nothingness. It is the arena for everything that happens: it has, latent in it, all the forces and particles that govern our physical world. Some physicists suspect that space can exist in different "phases," rather as water can exist in three forms: ice, liquid, and steam. Moreover, the present vacuum could be fragile and unstable. The analogy here is with water that is "supercooled." Water can cool below its normal freezing point if it is very pure and still; however, it takes only a small localised disturbance—for instance, a speck of dust falling into it—to trigger supercooled water's conversion into ice. Likewise, some have speculated that the concentrated energy created when particles crash together could trigger a "phase transition" that would rip the fabric of space itself. The boundary of the new-style vacuum would spread like an expanding bubble. In that bubble atoms could not exist: it would be "curtains" for us, for Earth, and indeed for the wider cosmos; eventually, the entire galaxy, and beyond, would be engulfed. And we would never see this disaster coming. The "bubble" of new vacuum advances as fast as light, and so no signal could forewarn us of our fate. This would be a cosmic calamity, not just a terrestrial one.

These scenarios may seem bizarre, but physicists discuss them with a straight face. The most favoured theories are reassuring: they imply that the risk is zero. But we cannot be one hundred percent sure what might actually happen. Physicists can dream up alternative theories (and even write down equations) that are consistent with everything we know, and therefore cannot be absolutely ruled out, but that would allow one or other of these catastrophes to happen. These alternative theories may not be frontrunners, but are they all so incredible that we needn't worry?

Back in 1983, physicists were already becoming interested in high-energy experiments of this kind. While visiting the Institute for Advanced Study in Princeton, I discussed these issues with a Dutch colleague, Piet Hut, who was also visiting Princeton and subsequently became a professor there. (The academic style of this institute, where Freeman Dyson has long been a professor, encourages "out of the box" thinking and speculations.) Hut and I realised that one way of checking whether an experiment is safe would be to see whether nature has already done it for us. It turned out that collisions similar to those being planned by the 1983 experimenters were a common occurrence in the universe. The entire cosmos is pervaded by particles known as cosmic rays that hurtle through space at almost the speed of light; these particles routinely crash into other atomic nuclei in space, with even greater violence than could be achieved in any currently feasible experiment. Hut and I concluded that empty space cannot be so fragile that it can be ripped apart by anything that physicists could do in their accelerator experiments. If it were, then the universe would not have lasted long enough for us to be here at all. However, if these accelerators became a hundred times more powerful—something that financial constraints still preclude, but which may be affordable if clever new designs are developed—then these concerns would revive, unless in the meantime our understanding has advanced enough to allow us to make firmer and more reassuring predictions from theory alone.

The old fears resurfaced more recently when plans were announced, both at the Brookhaven National Laboratory in the US and at the CERN laboratory in Geneva, to crash atoms together even more forcefully than had been done before. The director of the Brookhaven Laboratory at the time, John Marburger (now President Bush's scientific advisor), asked a group of experts to look into the issue. They did a calculation along the lines of the one that Hut and I had given, and offered reassurance that there was no threat of a cosmic Doomsday triggered by tearing the fabric of space.

But these physicists could not be quite so reassuring about the risk from strangelets. Collisions with the same energy certainly occur in the cosmos, but under conditions that differ in relevant respects from those of the planned terrestrial experiments; these differences could alter the likelihood of a runaway process.

Most of the "natural" cosmic collisions happen in interstellar space, in an environment so rarefied that even if they produced a strangelet, it would be unlikely to encounter a third nucleus, so there would be no chance of a runaway process. Collisions with Earth also differ in an essential way from those in accelerators, because incoming nuclei are stopped in the atmosphere, which does not contain heavy atoms like lead and gold.

Some fast-moving nuclei, however, impact directly on the Moon's solid surface, which does contain such atoms. Such impacts have occurred over its entire history. The Moon is nonetheless still there, and the authors of the Brookhaven report proffered this indisputable fact as reassurance that the proposed experiment couldn't wipe us out. But even these impacts differ in one possibly important way from those that would occur in the Brookhaven accelerator. When a fast particle crashes onto the Moon's surface, it hits a nucleus that is almost at rest, and gives it a "kick" or recoil. The resultant strangelets, produced as debris in the collision, would share this recoil motion, and therefore be sent hurtling through the lunar material. In contrast, the accelerator experiments involve symmetrical collisions, where two particles approach each other "head on." There is then no recoil: the strangelets have no net motion and therefore might stand more chance of grabbing ambient material.

Since the experiment would generate conditions that have never happened naturally, the only reassurance came from two theoretical arguments. First, even if strangelets could exist, theorists thought it unlikely that they would form in these violent collisions: it seemed more likely that the debris would disperse in the aftermath of the collision, rather than reassembling into a single lump. Second, if strangelets form, theorists would expect them to have a positive electric charge. On the other hand, to trigger runaway growth the strangelets would have to be negatively charged (so that they would attract, rather than repel, positively charged atomic nuclei in their surroundings).

The best theoretical guesses are therefore reassuring. Sheldon Glashow, a theorist, and Richard Wilson, an expert on energy and environmental issues, succinctly summarised the situation like this: "If strangelets exist (which is conceivable), and if they form reasonably stable lumps (which is unlikely), and if they are negatively charged (although the theory strongly favours positive charges), and if tiny strangelets can be created at the [Brookhaven] Relativistic Heavy Ion Collider (which is exceedingly unlikely), then there might just be a problem. A new-born strangelet could engulf atomic nuclei, growing relentlessly and ultimately consuming the entire Earth. The word 'unlikely,' however many times it is repeated, just isn't enough to assuage our fears of this total disaster."

What Risks Are Acceptable?

The accelerator experiments didn't give me any sleepless nights. Indeed, I don't know of any physicist who betrayed the slightest anxiety about them. However, these attitudes are little more than subjective assessments, based on some knowledge of the relevant science. The theoretical arguments depend on probabilities rather than certainties, as Glashow and Wilson spell out clearly. There is no evidence that exactly the same conditions have ever occurred naturally. We cannot be absolutely sure that strangelets couldn't lead to a runaway disaster.

The Brookhaven report (and a parallel effort by scientists from the biggest European accelerator, CERN, in Geneva) were presented as reassuring. However, even if one accepted their reasoning completely, the level of confidence they offered hardly seemed enough. They estimated that if the experiment were run for ten years, the risk of a catastrophe was no more than one in fifty million. These might seem impressive odds: a chance of disaster smaller than the chance of winning the UK's national lottery with a single ticket, which is about one in fourteen million. However, if the downside is destruction of the world's population, and the benefit is only to "pure" science, this isn't good enough. The natural way to measure the gravity of a threat is to multiply its probability by the number of people at risk, to calculate the "expected number" of deaths. The entire world's population would be at risk, so the experts were telling us that the expected number of human deaths (in that technical sense of "expected") could be as high as 120 (the number obtained by taking the world's population to be six billion and dividing by fifty million).

Obviously, nobody would argue in favour of doing a physics experiment knowing that its "fallout" could kill up to 120 people. This is not, of course, quite what we were told in this case: we were told instead that there could be up to one chance in fifty million of killing six billion people . Is this prospect any more acceptable? Most of us would, I think, still be uneasy. We are more tolerant of risks that we expose ourselves to voluntarily, or if we see some compensating benefit. Neither of these conditions pertains here (except for those physicists who are actually interested in what might be learnt from the experiment).

My Cambridge colleague Adrian Kent has emphasised a second factor: the finality and completeness of the extinction that this scenario would entail. It would deprive us of an expectation—important to most of us—that some biological or cultural legacy will survive our deaths; it would dash the hope that our lives and our works may be part of some continuing progress. It would, worse still, foreclose the existence of a (perhaps far larger) total number of people in all future generations. Wiping out all the world's people (and indeed destroying not just humans but the entire biosphere) could therefore be deemed far more than six billion times worse than the death of one person. So perhaps we should set an even more stringent threshold on the possible risk before sanctioning such experiments.

Philosophers have long debated how to balance the rights and interests of "possible people," who might have some future existence, against those of people who actually exist. For some, like Schopenhauer, the painless elimination of the world would not rate as an evil at all. But most would resonate more with Jonathan Schell's response: "While it is true that extinction cannot be felt by those whose fate it is—the unborn, who would stay unborn—the same cannot be said, of course, for extinction's alternative, survival. If we shut the unborn out of life, they will never have the chance to lament their fate, but if we let them into life they will have abundant opportunity to be glad that they were born instead of having been prenatally severed from existence by us. What we must desire first of all is that people be born, for their own sakes, and not for any other reason. Everything else—our wish to serve the future generations by preparing a decent world for them to live in, and our wish to lead a decent life ourselves in a common world made secure by the safety of the future generations—flows from this commitment. Life comes first, the rest is secondary."

Who Should Decide?

No decision to go ahead with an experiment with a conceivable "Doomsday downside" should be made unless the general public (or a representative group of them) is satisfied that the risk is below what they collectively regard as an acceptable threshold. The theorists in this episode seemed to have aimed to reassure the public about a concern that they considered unreasonable, rather than to make an objective analysis. The public is entitled to more safeguards than that. It isn't good enough to make a slapdash estimate of even the tiniest risk of destroying the world.

Francesco Calogero is one of the few who have addressed this issue thoughtfully. He is not only a physicist, but also a long-time activist for arms control, and a former general secretary of the Pugwash conferences. He expresses his concerns like this: "I am somewhat disturbed by what I perceive to be the lack of candour in discussing these matters. . . Many, indeed most [of those with whom I have had private discussion and exchanged messages] seem more concerned with the public relations impact of what they, or others, say or write, than in making sure that the facts are presented with complete scientific objectivity."

How should society guard against being unknowingly exposed to a not-quite-zero risk of an event with an almost infinite downside? Calogero suggests that no experiment that could conceivably carry such risks should be approved without a prior exercise, of a kind familiar from risk analyses in other contexts, involving a "Red Team" of experts (which would not include any of the group actually proposing the experiment) that would play devil's advocate, trying to think of the worst that might happen, and a "Blue Team" that would try to think of antidotes or counter-arguments.

When the purpose is to probe conditions where the physics is "extreme" and very poorly known, it is hard to rule out anything completely. Can we ever be sure enough of our reasoning to offer reassurance with the confidence level of a million, a billion, or even a trillion to one? Theoretical arguments can seldom offer adequate comfort at this level: they can never be firmer than the assumptions on which they are based, and only recklessly overconfident theorists would stake odds of a billion to one on the validity of their assumptions.

Even if a believable number could be assigned to the probability of a catastrophic outcome, the question remains: How low would the alleged risk have to be before we would give our informed consent to these experiments? There is no specific countervailing benefit to the rest of us, so the level would surely be lower than the experimenters might willingly accept on their own behalf. (It would also be far lower than the risk of nuclear devastation that citizens might have accepted during the Cold War, based on their personal assessment of what was at stake.) Some would argue that one chance in fifty million was low enough, because that is below the chance that within the next year an asteroid large enough to cause global devastation will strike Earth. (This is like arguing that the extra carcinogenic effect of artificial radiation is acceptable if it does not do more than double the risk from natural radiation.) But even this limit doesn't seem stringent enough. We may become resigned to a natural risk (like asteroids or natural pollutants) that we cannot do much about, but that doesn't mean that we should acquiesce in an extra avoidable risk of the same magnitude. Indeed, efforts are made to reduce risks far below that level whenever we can. That is why, for example, it is worth some effort to ameliorate the risk of asteroid impact.

UK government guidelines on radiation hazards deem it unacceptable that even the limited group of workers in a nuclear power station should risk more than one chance in one hundred thousand per year of dying through the effects of radiation exposure. If this very risk-averse criterion were applied to the accelerator experiment, taking the world's population as being at risk but accepting an equally stringent maximum number of deaths, we would require an assurance that the chance of catastrophe was below one in a thousand trillion (10^{-15}). If equal weight were attached to the lives of all potential people who might ever exist—a philosophically controversial stance, of course—then it could even be argued that the tolerable risk was up to a million times lower still.

The Hidden Cost of Saying No

This leads to a quandary. The most extreme precautionary policy would prohibit any experiment that created novel artificial conditions (unless we knew that the same conditions had already been created naturally somewhere). But this would utterly paralyse science. Obviously, producing a new kind of material—a new chemical, for example—shouldn't be banned: we are overwhelmingly confident that in such a case we understand the basic principles. But once we get to the threshold of danger, when the creation is, say, a new pathogen, then maybe we should pause. And physics experiments at ultrahigh energies break atomic nuclei into constituents that are not well understood, and so perhaps we should pause here, too.

There is a penumbra of cases in which, were we to put the clock back, caution might have been called for. For instance, refrigerators in scientific laboratories routinely use liquid helium to create temperatures within a fraction of a degree of absolute zero (-273 degrees centigrade). Nowhere in nature—not on Earth, nor even (we believe) elsewhere in the universe—is as cold as this: everything is warmed to nearly three degrees above absolute zero by the weak microwaves that are a relic of the universe's hot dense beginning, the afterglow of creation. Dr. Peter Michelson, of Stanford University, built a detector for cosmic gravitational waves, the slight ripples in the structure of space itself that astronomers predict should be generated by cosmic explosions. This instrument consisted of a metal bar, weighing over a tonne, cooled close to absolute zero in order to reduce the heat vibrations. He described this bar as "the coldest large object in the universe, not just on Earth." This boast may have been accurate (unless extraterrestrials had done similar experiments).

Should we really have worried when the first liquid-helium refrigerator was turned on? I think we should have. It is true that there were no theories at the time that pointed towards any danger. But that could have just been lack of imagination: there are some current theories (admittedly very unlikely ones) that predict a genuine risk, but when ultralow temperatures were first achieved, the uncertainties were far greater, and physicists surely couldn't have confidently claimed that the probability of catastrophe was less than one in a trillion. You might offer such extreme odds against the possibility that the Sun won't rise tomorrow, or that a fair die will give one hundred sixes in a row. But these cases depend on physical and mathematical principles that are readily understood and firmly "battle tested."

In deciding whether to sanction some new tampering with our environment, we need to ask: Is there really a deep and firm enough understanding that we can rule out catastrophe with a confidence level that reassures us? One cannot disagree with Adrian Kent's comment: "It is obviously unsatisfactory that the question of what constitutes an acceptable catastrophe risk should be decided, in an ad hoc way, according to the personal risk criteria of those who happen to be consulted—those criteria, however sincerely held and thoughtfully constructed, may be unrepresentative of general opinion."

Procedures with no specific aim beyond achieving a better understanding of nature and satisfying our curiosity should meet very stringent safety requirements. But we might acquiesce in riskier decisions being taken on our behalf if there were some compensating benefit, especially a large and urgent one. For instance, shortening the Second World War was almost certainly in the minds of Hans Bethe and Edward Teller when they calculated whether the first atomic bomb test would incinerate the entire atmosphere. With so much at stake, they might properly have gone ahead even without the ultrahigh level of reassurance that we would expect before sanctioning a peacetime academic experiment.

The accelerator experiments highlight a dilemma that will confront us more and more often in other sciences: who should decide (and how) whether a novel experiment should go ahead if a disastrous outcome is conceivable but believed to be very, very unlikely? They provide an interesting "test case" that forces us to focus—in a far more extreme context than any biological experiment—on how to assess asymmetric situations where the outcome will very probably be useful and positive, but could conceivably (but very improbably) be utterly disastrous. The Australian mousepox episode discussed earlier showed in microcosm what could happen if, even quite unintentionally, a dangerous pathogen were created and released. Later in the century nonbiological micromachines may be as potentially hazardous as rogue viruses, and an extreme

Drexler-style "grey goo scenario" may no longer seem like science fiction.

The "downside" from even the worst conceivable biological experiment would never be as bad as for the accelerator experiment, since the entire Earth would not be in jeopardy. But in the fields of biology and nanotechnology—in contrast to those that use huge particle accelerators—the experiments are smaller in scale, and so are likely to be done in far larger numbers, and in far greater variety. We then need assurance against even one of them going disastrously wrong. If a million separate experiments were to be conducted—a million chances of disaster—then the tolerable risk for each would be far lower than for a "one-shot" experiment. Quantifying these considerations into an actual number would require an estimate of the likely benefit. Greater risks would plainly be acceptable in experiments that were integral to a programme that could manifestly save millions of lives. The risks posed by science are sometimes the necessary concomitant of progress: if we don't accept some risk, we may forgo great benefits.

One special line of argument is used in risk assessment, with results that are often unduly optimistic. A major accident, for instance the destruction of an airliner or spacecraft, can occur in various different ways, each of which requires a whole series of mishaps (for instance, the combined or successive failure of several components). The pattern of risk can be expressed as a "fault tree"; the odds against each are then combined, rather as one multiplies the odds when betting on a combination of winners in horse racing (though the arithmetic is a bit more complicated because there may be several different failure modes, and the mishaps may be correlated in a way that the outcomes of separate horse races are not). Such calculations may overlook some crucial failure modes, and thereby offer a false sense of reassurance. The space shuttle was thought to be safe enough that the risk to its crew was less than one in a thousand. But the 1986 explosion happened on the twenty-fifth shuttle flight (and the tenth flight of the *Challenger* launch vehicle). In retrospect, odds of one in twenty-five would have been a better guess. Similarly, one should be cautious about the estimates given for various kinds of mishaps to nuclear power stations, which are calculated in a similar way.

To calibrate a tiny risk to the entire Earth, we multiply a very small probability by a colossal number, analogous to the most extreme asteroid impact events on the Torino scale. The probability is never quite zero because our fundamental knowledge of basic physics is incomplete; but even if it were very small indeed, when multiplied by a colossal number the product could still be big enough to worry about.

When a potentially calamitous downside is conceivable—not just in accelerator experiments, but in genetics, robotics, and nanotechnology—can scientists provide the ultraconfident assurance that the public may demand? What should be the guidelines for such experiments, and who should formulate them? Above all, even if guidelines are agreed upon, how can they be enforced? As the power of science grows, such risks will, I believe, become more varied and widely diffused. Even if each risk is small, they could mount up to a substantial cumulative danger.

10

THE DOOMSDAY
PHILOSOPHERS

---

*Can pure thought tell us whether humanity's
years are numbered?*

PHILOSOPHERS SOMETIMES ADVANCE ingenious arguments that may seem clinching to some, but to others seem
mere wordplay, or an intellectual sleight of hand, though it is
not easy to pinpoint the flaw. There is a modern philosophical
argument that humanity's future is bleak that may seem in this
dubious category, but which (with provisos) has weathered a
good deal of scrutiny. The argument was invented my friend
and colleague Brandon Carter, a pioneer in the use of the so-
called anthropic principle in science, the idea that laws govern-
ing the universe must have been rather special in order for life
and complexity to have emerged. He first presented this argu-

ment, to the bemusement of his academic audience, at a conference hosted by the Royal Society in London in 1983. The idea was actually just an afterthought in a lecture discussing the likelihood that life would evolve on planets orbiting other stars. It led Carter to conclude that intelligent life was rare elsewhere in the universe, and that even though the Sun would keep shining for billions of years, life's long-term future was bleak.

This "Doomsday argument" depends on a kind of "Copernican principle" or "principle of mediocrity" applied to our position in time. Ever since Copernicus, we have denied ourselves a central location in the universe. Likewise, according to Carter, we shouldn't assume that we are living at a special time in the history of humanity, neither among the very first nor among the very last of our species. Consider our place in the "roll call" of *Homo sapiens*. We know our place only very roughly: most estimates suggest that the number of human beings who have preceded us is around sixty billion, so our number in the roll call is in this range. A consequence of this figure is that ten percent of the people who have ever lived are alive today. At first sight this seems a remarkably high proportion, given that mankind can be traced back through thousands of generations. But for most of human history—the entire preagricultural era before (maybe) 8000 B.C.E.—there were probably fewer than ten million people in the world. By Roman times, world population was around three hundred million, and only in the nineteenth century did it rise above a billion. The dead outnumber the living, but only by a factor of ten.

Now consider two different scenarios for humanity's future: a "pessimistic" one, where our species dies out within one or two centuries (or if it survives longer than that has a much diminished population), so that the total number of humans who will ever exist is one hundred billion; and an "optimistic" scenario, where humanity survives for many millennia with at least the present population (or perhaps even spreads far beyond Earth with an ever-enlarging population), so that trillions of people are destined to be born in future. Brandon Carter argues that the "principle of mediocrity" should lead us to bet on the "pessimistic" scenario. Our place in the roll call (about halfway through) is then entirely unsurprising and typical, whereas in the "optimistic" scenario, where a high population persists into the far future, those living in the twenty-first century would be early in the roll call of humanity.

A simple analogy brings out the essence of the argument. Suppose that you are shown two identical urns: you are told that one contains just ten tickets, numbered from 1 to 10, and the other contains a thousand tickets, numbered from 1 to 1000. Suppose you pick one of the urns, draw a ticket from it, and find that you have drawn the number 6. You would then surely guess that you had, very probably, picked from the urn containing only 10 tickets: it would be very surprising to draw a ticket number as small as 6 from the urn containing a thousand tickets. Indeed, if you were equally likely, a priori, to have picked either urn, a simple probability argument then shows that having drawn the number 6, the odds are now one hundred to one that you actually chose from the urn containing only ten tickets.

Carter argues, along the same lines as in the case of the two urns, that our known place in the roll call of humans (about sixty billion human beings have preceded us) tilts the argument in favour of the hypothesis that there will be only one hundred billion humans, and would disfavour an alternative supposition that there would be more than one hundred trillion. So the argument suggests that the world's population cannot continue for many generation at its present level; either it must decline gradually, and be sustained at a far lower level than at present, or a catastrophe will overcome our species within a few generations.

An even simpler argument was used by Richard Gott, a professor at Princeton University with a thirty-year record of zany but original insights on faster-than-light travel, time machines, and the like. If we come upon some object or phenomenon, we are unlikely to be doing so very near the beginning of its life, nor very near its end. So it is a fair presumption that something that is already ancient will last for a long time in the future, and something that is of recent origin shouldn't be expected to be so durable. Gott recalls, for instance, that in 1970 he visited the Berlin Wall (then twelve years old) and the pyramids (over four thousand years old); his argument would have predicted (correctly) that the pyramids would be very probably be still standing in the twenty-first century; but it would be unsurprising if the Berlin Wall wasn't (and of course, it has gone).

Gott even showed how the argument applied to Broadway shows. He made a list of all the plays and musicals that were running on Broadway on a particular day (May 27, 1993) and found out how long each show had been open. On that basis, he predicted that those that had been running longest would survive furthest into the future. *Cats* had already been running for 10.6 years, and it kept going for more than seven years more. Most of the others, which had been running less than a month, closed within a few more weeks.

Of course, most of us could have made all Gott's predictions without using his line of argument at all, from our familiarity with basic history, the general robustness and durability of artefacts of different types, and so forth. We also know about American tastes, and the economics of the theatre. The more background information we have, the more confident can our predictions be. But even a newly landed alien devoid of such background knowledge, who knew nothing except how long these various phenomena had existed, could have used Gott's argument to make some crude but correct predictions. And of course the future duration of humanity is something about which we are as ignorant as any Martian would be about the sociology of Broadway shows. Gott therefore argues, following Carter, that this line of reasoning can tell us something—indeed, something far from cheerful—about the likely longevity of our species.

Obviously humankind's future cannot be stripped down to a simple mathematical model. Our destiny depends on multitudinous factors, above all—a main theme of this book—on choices that we ourselves make during the present century. The Canadian philosopher John Leslie takes the line that the Doomsday argument nonetheless tilts the odds: it should make you less optimistic about humanity's long-range future than you would otherwise be. If you thought, a priori, that it was overwhelmingly probable that humanity would continue, with a high population, for millennia, then the Doomsday argument would reduce your confidence, though you might still end up favouring that scenario. This can be understood by generalising the urn example. Suppose that instead of just two urns, there were millions of urns that each contained a thousand tickets and only one that contained just ten. Then if you pick an urn at random, you would be surprised to draw a 6. But if there were millions of "thousand-ticket" urns, then it would be less surprising that you had drawn an unusually low number from one of them than that you had picked the unique urn with only ten tickets in it. Likewise, if the a priori probability strongly favours a prolonged future for humanity, then "doom soon" might be less likely than finding ourselves coming very early in the roll call of humanity.

Leslie can thereby resolve another conundrum that at first sight seems to discredit the entire line of argument. Suppose that we had a fateful decision that would determine whether the species might soon be extinguished, or else whether it would survive almost indefinitely. For instance, this might be the choice of whether to foster the first community away from Earth, which, once established, would spawn so many others that one would be guaranteed to survive. If such a community were indeed established and flourished, we would currently find ourselves exceedingly early in the roll call. Does the Doomsday argument somehow constrain us towards the choice that leads to a truncated human future? Leslie argues that we are free to choose, but that the choice we make affects the prior probability of the two scenarios.

Another ambiguity concerns who or what should be counted: How do we define humanity? If the entire biosphere were to be wiped out in some global catastrophe, then there is no doubt about when the roll call ends. But if our species were to morph into something else, would this amount to the end of humanity? If so, the Carter–Gott argument might be telling us something different: it could offer support for Kurzweil, Moravec, and others who predict a "takeover" by machines within this crucial century. Or suppose there are other beings on other worlds. Then perhaps all intelligent beings, not just humans, should be in the "reference class." There is then no clear way to order the roll call, and the argument collapses. (Gott and Leslie have used similar reasoning to argue against there being other worlds with much higher populations than ours. If there were, they claim, we should be surprised not to be in one of them.)

When I first heard Carter's Doomsday argument, it reminded me of George Orwell's robust comment in a different context: "You must be a real intellectual to believe that—no ordinary person could be so foolish." But pinpointing an explicit flaw is not a trivial exercise. It is worth doing so, however, since none of us welcomes a new argument that humanity's days may be numbered.

11

THE END OF SCIENCE?

---

*Future Einsteins may transcend current
theories of space, time and microworld.
But the holistic sciences of life and complexity
pose mysteries that human minds may never
fully grasp.*

WILL SCIENCE CONTINUE TO SURGE FORWARD, bringing new insights, and perhaps further threats as well? Or will the science of the coming century be an anticlimax after the triumphs already achieved?

The journalist John Horgan has claimed the latter: he argues that we have already uncovered all the really big ideas. All that remains, according to Horgan, is to fill in the details, or else to indulge in what he terms "ironic science"—flaky, ill-disciplined conjectures about topics that will never come within the ambit of serious empirical study. I believe that this thesis is fundamentally mistaken, and that ideas as revolutionary as any that were discovered in the twentieth century remain to be disclosed. I prefer Isaac Asimov's viewpoint. He likened science's frontier to a fractal—a pattern with layer upon layer of structure, so that a tiny bit, when magnified, is a simulacrum of the whole: "No matter how much we learn, whatever is left, however small it may seem, is just as infinitely complex as the whole was to start with."

Twentieth-century advances in understanding atoms, life, and the cosmos rank as humankind's greatest collective intellectual achievement. (The proviso "collective" is crucial. Modern science is a cumulative enterprise; discoveries are made when the time is ripe, when the key ideas are "in the air," or when some novel technique is exploited. Scientists aren't quite as interchangeable as light bulbs, but there are nonetheless few cases in which an individual has made much difference to the long-run development of the subject: if "A" hadn't done the work or made the discovery, "B" would before long have done something similar. This is the way science normally develops. A scientist's work loses its individuality, but it lasts. Einstein has a specially honoured place in the scientific pantheon because he was one of the few exceptions: had he not existed, his deepest insights would have emerged much later, perhaps by a different route and through the efforts of several people rather than just one. But the insights would eventually have been achieved: not even Einstein left a distinctive personal imprint to match that of the greatest writers or composers.)

Ever since the classical Greek era when earth, air, fire, and water were believed to be the substances of the world, scientists have sought a "unified" picture of all the basic forces of nature, and to understand the mystery of space itself. Cosmologists are sometimes berated for being "often in error but never in doubt." They have indeed often embraced poorly grounded speculations with irrational fervour, and been led by wishful thinking to read too much into vague and tentative evidence. But even the more cautious among us are confident that we have now grasped at least the outlines of our entire cosmos, and learnt what it is made of. We can trace the evolutionary story back before our solar system formed—indeed, back to an epoch long before there were any stars, when everything sprouted from an intensely hot "genesis event," the so-called big bang, about fourteen billion years ago. The first microsecond is shrouded in mystery, but everything that happened since then—the emergence of our complex cosmos from simple beginnings—is the outcome of laws that we can understand, even though the details still elude us. Just as geophysicists have come to understand the processes that made the oceans and sculpted the continents, so astrophysicists can understand our Sun and its planets, and indeed the other planets that may orbit distant stars.

In earlier centuries, navigators mapped the outlines of the continents and took the measure of Earth. Within just the last few years our map of the cosmos, in time and in space, has likewise firmed up. A challenge for the twenty-first century is to refine our present picture, filling in ever more detail, just as generations of surveyors did for Earth, and especially to probe the mysterious domains where earlier cartographers wrote "here be dragons."

Shifting Paradigms

The term "paradigm" was popularised by Thomas Kuhn in his classic book *The Structure of Scientific Revolutions*. A paradigm is not just a new idea (if it were, most scientists could claim to have shifted a few): a paradigm shift denotes an intellectual upheaval that reveals new insights and transforms our scientific perspective. The greatest paradigm shift of the twentieth century was the quantum theory. This theory tell us, completely counter to all intuition, that on the atomic scale nature is intrinsically "fuzzy." Nonetheless, atoms behave in precise mathematical ways when they emit and absorb light, or link together to make molecules. A hundred years ago, the very existence of atoms was controversial; but quantum theory now accounts for almost every detail of how atoms behave. As Stephen Hawking points out, "It is a tribute to how far we have come in theoretical physics that it now takes enormous machines and a great deal of money to perform an experiment [on subatomic particles] whose result we cannot predict."

Quantum theory is vindicated every time you take a digital photograph, surf the Internet, or use any gadget—a CD player, or a supermarket bar code—that involves a laser. Even now some of its astonishing implications are just dawning on us. It may allow computers to be designed on entirely new principles, that could outperform any "classical" computer, however long Moore's law continues.

Another new paradigm of twentieth-century science—another astonishing intellectual leap—is largely the creation of one man, Albert Einstein: he deepened our understanding of space, time, and gravity, giving us a theory, general relativity, that governs the motions of planets, stars, and the expanding universe itself. This theory is now confirmed by very precise radar tracking of planets and spacecraft, and by astronomical studies of neutron stars and black holes—objects where gravity is so strong that space and time are grossly distorted. Einstein's theory might have seemed arcane, but it is vindicated every time a truck or plane fixes its position via the global positioning satellite (GPS) system.

Linking the Very Large and the Very Small

But Einstein's theory is inherently incomplete: it treats space and time as a smooth continuum. If we chop a piece of metal (or indeed, any material at all) into smaller and smaller pieces, there is an eventual limit when we reach the quantum level of individual atoms. Likewise, on the very tiniest scale, we expect even space itself to be grainy. Perhaps not just space, but also time itself, is made up of finite quanta rather than "flowing" continuously. There may be a fundamental limit to how precisely any clock can ever subdivide time. But neither Einstein's theory nor quantum theory, in their present forms, can tell us about the microstructure of space and time. Twentieth-century science left this major piece of unfinished business as a challenge for the twenty-first.

The history of science suggests that when a theory breaks down, or confronts a paradox, the resolution will be a new paradigm that transcends what went before. Einstein's theory and the quantum theory cannot be meshed together: both are superb within limits, but at the deepest level they are contradictory. Until there has been a synthesis, we certainly will not be able to tackle the overwhelming question of what happened right at the very beginning, still less attach any meaning to the question, "What happened before the big bang?" At the "instant" of the big bang everything was squeezed smaller than a single atom, so quantum fluctuations could shake the entire universe.

According to superstring theory, the currently most favoured approach to a unified theory, the particles that make up atoms are all woven from space itself. The fundamental entities are not points, but tiny loops, or "strings," and the various subnuclear particles are different modes of vibration—different harmonics—of these strings. Moreover, these strings are vibrating not in our ordinary space (with three spatial dimensions, plus time) but in a space of ten or eleven dimensions.

Beyond Our Space and Time

We appear to ourselves as three-dimensional beings: we can go left or right, forward or backward, up or down, and that is all. So how are the extra dimensions, if they exist, concealed from us? It may be that they are all wrapped up tightly. A long hose pipe may look like a line (with just one dimension) when viewed from a distance, but from closer up we realise that it is a long cylinder (a two-dimensional surface) rolled up tightly; from still closer, we realise that this cylinder is made from material that is not infinitely thin, but extends in a third dimension. By analogy, every apparent point in our three-dimensional space, if hugely magnified, may actually have some complex structure: a tightly wound origami in several extra dimensions.

Some of the extra dimensions could conceivably show up on a microscopic scale in laboratory experiments (though they are probably wrapped too tightly even for that). Even more interestingly, one extra dimension may not be wrapped up at all: there may be another three-dimensional universe "alongside" ours, embedded in a grander-dimensional space. Bugs crawling around on a large sheet of paper (their two-dimensional "universe") may be unaware of a similar sheet that is parallel to it and not in contact. Likewise, there could be another entire universe (three-dimensional, like ours) less than a millimetre away from us, but we are oblivious to it because that millimetre is measured in a fourth spatial dimension, and we are imprisoned in just three.

There could have been many big bangs, even an infinity of

them, not just the one that led to "our" universe. Even our own "universe," the aftermath of our own big bang, may extend far beyond the ten billion light years that our telescopes can probe: it may encompass a still vaster domain, extending so far away that no light from it has yet had time to reach us. Whenever a black hole forms, processes deep inside it could perhaps trigger the creation of another universe, which would expand into a space disjoint from our own. If that new universe were like ours, then stars, galaxies, and black holes would form in it, and those black holes would in turn spawn another generation of universes, and so on, perhaps ad infinitum. Perhaps universes could be created in a futuristic laboratory, by imploding a lump of material to make a small black hole, or even by crashing together atoms boosted to very high energies in a particle accelerator. If so, the theological arguments from design could be resuscitated in a novel guise, blurring the boundary between the natural and the supernatural.

We have learnt, in the time since Copernicus dethroned Earth from its central position, that our solar system is just one of billions within range of our telescopes. Our cosmic horizons are now, once again, enlarging just as dramatically: what we have traditionally called our universe may be just one "island" in an infinite archipelago.

To make scientific predictions one needs to believe that nature is not capricious, and to have uncovered some regular patterns. But these patterns need not be fully understood. For example, the Babylonians, more than two thousand years ago, could predict when solar eclipses were likely, because they had already gathered data for centuries and discovered repetitive patterns in the timing of eclipses (in particular, that they follow an eighteen-year cycle). But the Babylonians did not know how the Sun and Moon actually moved. It was not until the seventeenth century—the era of Isaac Newton and Edmund Halley—

that the eighteen-year cycle was attributed to a "wobble" in the orbit of the Moon.

Quantum mechanics works marvellously: most scientists apply it almost unthinkingly. As my colleague John Polkinghorne has put it, "The average quantum mechanic is no more philosophical than the average motor mechanic." But many thoughtful scientists from Einstein onwards have found the theory "spooky" and doubt that we have yet attained the optimum perspective on it. Interpretations of quantum theory today may be on a "primitive" level, analogous to the Babylonian knowledge of eclipses: useful predictions, but no deep understanding.

Some of the baffling paradoxes of the quantum world may be clarified by an idea familiar from science fiction: "parallel universes." Olaf Stapledon's classic novel *Star Maker* prefigured this concept. The star maker is a creator of universes, and in one of his more sophisticated creations, "Whenever a creature was faced with several possible courses of action, it took them all, thereby creating many . . . distinct histories of the cosmos. Since in every evolutionary sequence of the cosmos there were many creatures and each was constantly faced with many possible courses, and the combinations of all their courses were innumerable, an infinity of distinct universes exfoliated from every moment."

At first sight, the concept of parallel universes might seem too arcane to have any practical impact. But it may actually offer the prospect of an entirely new kind of computer, the quantum computer, which can transcend the limits of even the fastest digital processor by, in effect, sharing the computational burden among a near infinity of parallel universes.

In the twentieth century we learnt the atomic nature of the entire material world. In the twenty-first, the challenge will be to understand the arena itself, to probe the deepest nature of space and time. New insights should clarify how our universe began, and whether it is one of many. On a more practical terrestrial level, they may reveal new sources of energy latent in empty space itself.

A fish may be barely aware of the medium in which it lives and swims; certainly, it has no intellectual powers to comprehend that water consists of interlinked atoms of hydrogen and oxygen, each made up of still smaller particles. The microstructure of empty space could, likewise, be far too complex for unaided human brains to grasp. Ideas on extra dimensions, string theory, and the like will attract lively scientific interest in this century. We aspire to understand our cosmic habitat—and unless we try, we certainly will not succeed—but it could be that we stand little more chance than a fish.

The Boundaries of Time

Time, as Wells and his chrononaut knew, is a fourth dimension. Time travel into the far future violates no fundamental physical laws. A spaceship that could travel at 99.99 percent of the speed of light would allow its crew to "fast forward" into the future. An astronaut who managed to navigate into the closest possible orbit around a rapidly spinning black hole without falling in could, in a subjectively short period, view an immensely long future time span in the external universe. Such adventures may be unfeasible, but they are not physically impossible.

But what about travel into the past? More than 50 years ago, the great logician Kurt Gödel invented a bizarre hypothetical universe, consistent with Einstein's theory, that allowed, "time loops," in which events in the future "cause" events in the past that then "cause" their own causes, introducing a lot of weirdness to the world but no contradictions. (The film, *The Terminator*, in which a son sends his father back in time to save (and inseminate) his mother, wonderfully combines the insights of the greatest Austrian-American mind, Gödel, with the talents of the greatest Austrian-American body, Arnold Schwarzenegger.) Several later theorists have used Einstein's theories to design "time machines" that might create temporal loops. But these are not machines that would fit in a Victorian basement. Some of them need to be of effectively infinite length; others need vast amounts of energy. Returning to the past involves the risk of changing it in such a way that makes history internally inconsistent by, for instance, preventing your parents from being born. Such conundrums do not rule out time travel even in principle: they merely restrict the time traveler's free will. But that is nothing new. Physics already constrains us: we cannot exercise our free will by walking on the ceiling. Another option is that time travelers could shift into a parallel universe, where events played out differently rather than repeating themselves, as in the film *Groundhog Day*.

We plainly do not yet have a unified theory; and parallel universes, time loops and extra dimensions are surely "big ideas" for twenty-first-century science. Acknowledging this, Horgan can only sustain his pessimistic "end of science" thesis by disparaging such theories as "ironic science." This is probably a fair assessment of their present status, when they are a set of mathematical ideas, laced with what seems science fiction and disengaged from experiment or observation. But the hope is that such theories, if within our intellectual grasp, will actually explain things about our physical world that now seem mysterious: why protons, electrons, and other subatomic particles actually exist, and why the physical world is governed by particular forces and laws. A unified theory may reveal some unsuspected things, either on tiny scales, or by explaining some mysteries of our expanding universe. Perhaps some novel form of energy latent in space can be usefully extracted; an understanding of extra dimensions could give substance to the concept of time travel. Such a theory will also tell us what kinds of extreme experiments, if any, could trigger catastrophe.

The Third Frontier of Science: The Very Complex

A definitive theory for cosmos and micro world—even if it were some day achieved—would still not presage the "end of science." There is another open frontier: the study of things that are very complicated—above all, ourselves and our habitat. We may understand an individual atom, and even the mysteries of the quarks and other particles that lurk within its nucleus, but we are still perplexed by the intricate way atoms combine to make all the elaborate structures in our environment, especially those that are alive. The phrase "theory of everything," often used in popular books, has connotations that are not only hubristic, but very misleading. A so-called theory of everything would actually offer absolutely zero help to ninety-nine percent of scientists.

The brilliant and charismatic physicist Richard Feynman liked to emphasise this point with a nice analogy, which actually dates back to T.H. Huxley in the nineteenth century. Imagine that you had never seen chess being played before. By watching a few games, you could infer the rules. But in chess, learning how the pieces move is just a trivial preliminary on the absorbing progression from novice to grandmaster. Likewise, even if we knew the basic laws, exploring how their consequences have unfolded over cosmic history—how galaxies and stars and planets formed, and how here on Earth, and perhaps in many biospheres elsewhere, atoms assembled into creatures able to reflect on their origins—is an unending challenge.

Science is still just beginning: each advance brings into focus a new set of questions. I agree with John Maddox that "The big surprises will be the answers to questions that we are not yet smart enough to ask. The scientific enterprise is an unfinished project and will remain so for the rest of time."

It may seem presumptuous for cosmologists to pronounce confidently on arcane and remote matters when the views of experts in long-studied everyday subjects such as diet and child-care are manifestly little more than transient fashions. Yet what makes things hard to understand is how complicated they are, not how big they are. Planets and stars are big, but move in accord with simple laws. We can understand stars, and atoms as well; but the everyday world, especially the living world, poses a greater challenge. Dietetics is, in a real sense, a harder science than cosmology or subatomic physics. Human beings, the most intricately constructed entities we are aware of in the universe, are midway between atoms and stars. It would take as many human bodies to make up the Sun as there are atoms in each of us.

Our everyday world poses a still greater challenge to twenty-first century science than either the cosmos or the world of subnuclear particles. The biological realm is the main challenge, but even simple substances behave in complex ways. Weather patterns are manifestations of the well-understood physics of air and water, but are exceedingly intricate, chaotic, and unpredictable; improved theories of the micro world are no help at all to weather forecasters.

When we grapple with the complexities on our human scale, a holistic approach proves more helpful than naive reductionism. Animal behaviour makes the most sense when understood in terms of goals and survival. We can predict with confidence that an albatross will return to its nesting place after wandering ten thousand kilometres or more. Such a prediction would be impossible—not just in practice, but even in principle—if we analyzed the albatross into an assemblage of electrons, protons, and neutrons.

The sciences are sometimes likened to different levels of a tall building: logic in the basement, mathematics on the first floor, then particle physics, then the rest of physics and chemistry, and so forth, all the way up to psychology, sociology, and economics in the penthouse. But the analogy is poor. The superstructures, the "higher-level" sciences dealing with complex systems, aren't imperilled by an insecure foundation, as a building is. There are laws of nature in the macroscopic domain that are just as much of a challenge as anything in the micro world, and are conceptually autonomous from it—for instance, those that describe the transition between regular and chaotic behaviour, which apply to phenomena as disparate as dripping water pipes and animal populations.

Problems in chemistry, biology, the environment, and human sciences remain unsolved because scientists haven't elucidated the patterns, structures, and interconnections, not because we don't understand subatomic physics well enough. In trying to understand how water waves break, and how insects behave, analysis at the atomic level doesn't help. Finding the "readout" of the human genome—discovering the string of molecules that encode our genetic inheritance—is an amazing achievement. But it is just the prelude to the far greater challenge of postgenomic science: understanding how the genetic code triggers the assembly of proteins and expresses itself in a developing embryo. Other aspects of biology, especially the nature of the brain, pose challenges that can barely yet be formulated.

The Limits of Human Minds

Some branches of science could one day come to a halt. But this may happen because we come up against limits of what our brains can understand, rather than because the subject is exhausted. Physicists may never understand the bedrock nature of space and time because the mathematics is just too hard; but I think our efforts to understand very complex systems—above all, our own brains—will be the first to hit such limits. Perhaps complex aggregates of atoms, whether brains or machines, can never understand everything about themselves.

Computers with human-level capabilities will accelerate science, even though they won't think the way we do. IBM's chess-playing computer Deep Blue didn't evolve its strategy like a human player; it exploited its computational speed to compare millions of alternative series of moves and responses, applying a complicated set of rules, before deciding on an optimum move. This "brute force" approach overwhelmed a world champion; likewise, machines will make scientific discoveries that have eluded unaided human brains. For example, some substances completely lose their electrical resistance when cooled to very low temperatures (superconductors). There is a continuing quest to find the "recipe" for a superconductor that works at ordinary room temperatures (that is, nearly three hundred degrees above absolute zero; the highest superconducting temperature achieved so far is 120 degrees). This quest involves a great deal of "trial and error," because nobody understands exactly what makes the electrical resistance disappear more readily in some materials than in others.

Suppose that a machine came up with such a recipe. It might have succeeded in the same way that Deep Blue won its chess games against Kasparov: by testing out millions of possibilities rather than by having a human-style theory or strategy. But it would have achieved something that would get a scientist a Nobel Prize. Moreover, its discovery would herald a technical breakthrough that could, among other things, lead to still more powerful computers, an example of the runaway acceleration in technology, worrying to Bill Joy and other futurists, that could be unstoppable when computers can augment or even supplant human brains.

Simulations, using ever more powerful computers, will help scientists to understand processes that we neither study in our laboratories nor observe directly. My colleagues can already create a "virtual universe" in a computer, and do "experiments" on it—simulating, for example, how stars form and die, and how our Moon formed in a crash between the young Earth and another planet.

The First Life

Soon, biologists will have clarified the processes whereby combinations of genes encode the intricate chemistry of a cell, and the morphology of limbs and eyes. Another challenge is to elucidate how life began, and perhaps even replicate the event, either in a laboratory or "virtually" in a computer (where evolution can be studied much faster than in real time).

All life on Earth seems to have had a common ancestor, but how did this first living thing come into being? What led from amino acids to the first replicating systems, and to the intricate protein chemistry of unicellular life? The answer to this question—the transition from the nonliving to the living—is fundamental unfinished business for science. Laboratory experiments that try to simulate the "soup" of chemicals on the young Earth may offer clues; so might computer simulations. Darwin envisaged a "warm little pond." We are now more aware of the immense variety of niches that life can occupy. The ecosystems near hot sulphurous outwellings in the deep oceans tell us that not even sunlight is essential. So life's beginnings may have occurred in a torrid volcano, a location deep underground, or even in the rich chemical mix of a dusty interstellar cloud.

Above all, we want to know whether life's emergence was in some sense inevitable, or whether it was a fluke. Our Earth's cosmic importance depends on whether biospheres are rare or common, which depends in turn on how "special" the conditions need to be for life to begin. The answer to this key question affects the way we see ourselves and Earth's long-range future. We are stymied, of course, by the fact that we have just a single example, but this may change. The quest for alien life is perhaps the most fascinating challenge for twenty-first-century science. Its outcome will influence our concept of our place in nature as profoundly as Darwinism has over the last 150 years.

12

DOES OUR FATE HAVE COSMIC SIGNIFICANCE?

The odds could be so heavily stacked against the emergence (and survival) of complex life that Earth is the unique abode of conscious intelligence in our entire Galaxy. Our fate would then have truly cosmic resonance.

IS LIFE WIDESPREAD? Or is Earth special—not just to us, for whom it is the home planet, but for the wider cosmos?

So long as we know about only one biosphere, our own, we cannot exclude its being unique: complex life could be the outcome of a chain of events so unlikely that it happened only once within the observable universe, on the planet where (of course) we are. On the other hand, life could be widespread, emerging on any Earth-like planet (and perhaps in many other cosmic environments too). We still know too little about how life began and how it evolves to decide between these two extreme possibilities. The greatest breakthrough would be to find another biosphere: real alien life.

Unmanned explorations of the solar system in the coming decades may firm up the odds. Since the 1960s, space probes have been sent to the other planets of our solar system, beaming back pictures of worlds that are varied and distinctive; but none—in sharp contrast to our own planet—seem hospitable to life. Mars is still the main focus of attention. Probes have revealed dramatic Martian landscapes: volcanoes up to twenty kilometres high, and a canyon six kilometres deep and stretching four thousand kilometres across the planet. There are dried-up river beds, even features that look like the shoreline of a lake. If surface water once flowed on Mars, it is likely to have originated deep underground, and been forced up through thick permafrost.

Probing Mars and Beyond

NASA's first serious search for Martian life was in the 1970s. The *Viking* probes parachuted onto a barren rock-strewn desert and scooped up samples of soil; their instruments detected no sign of even the most primitive organisms. The only serious claim for fossil life came later, from analyses of a piece of Mars that made its own way to Earth. Mars is being battered, as is Earth, by asteroid impacts that throw debris out into space. Some of this debris, after wandering in orbit for many million years, strikes Earth as meteorites. In 1996, NASA officials orchestrated a much-hyped press conference, even attended by President Clinton, to proclaim that a meteorite recovered from the Antarctic, with chemical signatures of Martian origin, carried traces of tiny organisms. Scientists have been backtracking ever since: "life on Mars" may vanish just as the "canals" did a century ago. But hope of life on the red planet has not been abandoned, though even the optimists expect little more than dormant bacteria. Further space probes will analyse the Martian surface far more thoroughly than *Viking* did, and (in later missions) return samples to Earth.

Mars is not the only target for these reconnoitres. In 2004 the European Space Agency's *Huygens* probe, part of the cargo of NASA's *Cassini* mission, will parachute into the atmosphere of Titan, Saturn's giant moon, seeking anything that might be alive. There are longer-term plans to land a submersible probe on Jupiter's moon Europa, to seek life—perhaps even with fins or tentacles—in its ice-covered oceans.

Detecting life in two places in our solar system—which we now know is just one of millions of planetary systems in our galaxy—would suggest that life is common elsewhere in the universe. We would immediately conclude that our universe (with billions of galaxies each containing billions of stars) could harbour trillions of habitats where some kind of life (or vestiges of past life) exist. That is why it is scientifically so important to search for life on the other planets and moons of our solar system.

There is one key proviso, however: before drawing any inference about the ubiquity of life, we would need to be quite sure that any extraterrestrial life had begun independently, and that organisms had not made their way, via cosmic dust or meteorites, from one planet to another. After all, we know that some meteorites that hit Earth have come from Mars; if there was life on them, maybe that is how life began on Earth. Perhaps we all have a Martian ancestry.

Other Earths?

Even if there is life elsewhere in our solar system, few if any scientists expect it to be "advanced." But what about the remoter cosmos? In the years since 1995, a new field of a science has opened up: the study of other families of planets, in orbit around distant stars. What are the prospects of life on some of these? Few of us were surprised that these planets existed: astronomers already knew that other stars formed as our Sun did, from a slowly spinning interstellar cloud that contracted into a disc; the dusty gas in these other discs could agglomerate into planets, just as happened around the new-born Sun. But until the 1990s there were no techniques sensitive enough actually to disclose any of these faraway planets. At the time of writing, a hundred other stars like the Sun are already known to have at least one planet; almost every month more are being discovered. Those planets found so far, orbiting solar-type stars, are all roughly the size of Jupiter or Saturn, the giants of our own solar system. But these are probably just the largest members of other "solar systems" whose smaller members remain to be discovered. A planet like Earth, three hundred times less massive than Jupiter, would be too small and faint be revealed by present techniques, even if it were orbiting one of the very nearest stars. To observe Earth-like planets will require very large telescope arrays in space. NASA's flagship science programme, "Origins," is focussed on the origin of the universe, of planets, and of life. One of its most exciting projects will be the so-called Terrestrial Planet Finder, an array of telescopes in space; Europeans are planning a similar project, called "Darwin."

We were all, when young, taught the layout of our own solar system—the sizes of the nine major planets, and how they move in orbit around the Sun. But twenty years from now, we will be able to tell our grandchildren far more interesting things on a starry night. Nearby stars will no longer just be twinkling dots in the sky. We will think of them as the Suns of other solar systems. We will know the orbits of each star's retinue of planets, and even some topographic details of the bigger planets.

The Terrestrial Planet Finder and its European counterpart should discover many such planets, but only as faint points of light. Nonetheless, much can be learnt about them even without a detailed picture. Viewed from (say) fifty light years away—the distance of a nearby star—Earth would be, in Carl Sagan's phrase, a "pale blue dot," seeming very close to a star (our Sun) that outshines it by a factor of many billions. The shade of blue would be slightly different, depending on whether the Pacific ocean or the Eurasian land mass was facing us. By observing other planets, even if we can't resolve detail on their surfaces, we can therefore infer whether they are spinning, the length of their "day," and even their gross topography and climate.

We will be especially interested in possible "twins" of our Earth: planets the same size as ours, orbiting other Sun-like stars, and with temperate climates where water neither boils nor stays frozen. By analysing such a planet's faint light, we could infer what gases existed in its atmosphere. If ozone existed—implying that it was rich in oxygen, as our Earth's atmosphere is—this would indicate a biosphere. Our own atmosphere didn't start out that way, but was transformed by primitive bacteria in its early history.

But an actual image of such a planet—one that can be displayed on the wall-sized screens that will by then have replaced posters as room decorations—will surely have even more impact than the classic pictures of our own planet viewed from space. Even if NASA-type programmes continued for several decades, we won't have such pictures until after 2025. They will require huge mirrors in space; even an array spread over hundreds of kilometres would give a very blurred and crude image, just able to reveal an ocean or a continental land mass. Still further ahead, robotic fabricators may be able to construct, in the zero gravity of space, gossamer-thin mirrors on an even more gigantic scale. These would show more detail, and allow us to probe even further away, increasing the chance of finding a planet that might harbour life.

Alien Life?

How far away will we have to search to find another biosphere? Does life start on every planet in the right temperature range, where there is water, along with other elements such as carbon? At present, such questions are open. As often in science, lack of evidence leads to polarised and often dogmatic opinions, but agnosticism is really the only rational attitude while we know so little about how life began, how varied its forms and habitats could be, and what evolutionary paths it might take.

Could some of these planets, orbiting other stars, harbour life forms far more exotic than even optimists might expect on Mars or Europa—even something that could be called intelligent? To firm up the odds we need a clearer understanding of just how special Earth's physical environment had to be in order to permit the prolonged selection process that led to the higher animal forms on Earth. Donald Brownlee and Peter Ward, in their book *Rare Earth*, claim that very few planets around other stars—even those that resembled Earth in their size and temperatures—would provide the requisite long-term stability for the prolonged evolution that must precede advanced life. They think that there are several other prerequisites, which might be fulfilled only rarely. The planet's orbit must not wander too close to its "sun," nor too far away, as it would if other larger planets came too close and nudged it into a different orbit; its spin must be stable (something that depends on our Moon being large); there must not be excessive bombardment by asteroids; and so forth.

But the greatest uncertainties lie in the province of biology, not astronomy. First, how did life begin? I think that there is a real chance of progress here, so that we will know whether it is a "fluke," or whether it is nearly inevitable in the kind of initial "soup" expected on a young planet. But there is a second question: Even if simple life exists, what are the odds against it evolving into something that we would recognise as intelligent? This is likely to prove far more intractable. Even if primitive life were common, the emergence of "advanced" life may not be.

We know, in outline, the key stages in life's development here on Earth. The simplest organisms seem to have emerged within one hundred million years of the final cooling of Earth's crust after the last major impact, about four billion years ago. But about two billion years seems to have elapsed before the first eukaryotic (nucleated) cells appeared, and a further billion before multicellular life. Most of the standard body types seem to have first appeared during the "Cambrian explosion" just over half a billion years ago. The immense variety of creatures on land emerged since that time, punctuated by major extinctions, such as the event sixty-five million years ago that wiped out the dinosaurs.

Even if simple life existed on many planets around nearby stars, complex biospheres like Earth's could be rare: there could be some key hurdle in evolution that is hard to surmount. Perhaps it is the transition to multicellular life. (The fact that simple life on Earth seems to have emerged quite quickly, whereas even the most basic multicellular organisms took nearly three billion years, suggests that there may be severe barriers to the emergence of any complex life.) Or the biggest hurdle could come later. Even in a complex biosphere the emergence of human-level intelligence isn't guaranteed. If, for instance, the dinosaurs hadn't been wiped out, the chain of mammalian evolution that led to Homo sapiens may have been foreclosed, and we cannot predict whether another species would have taken our role. Some evolutionists regard the emergence of intelligence as a contingency, even an unlikely one. Others dissent from this line, however. In the latter camp is my Cambridge colleague Simon Conway Morris, an authority on the extraordinary variety of Cambrian life forms in the Burgess Shale, in the Canadian Rockies in British Columbia. He is impressed by the evidence for "convergence" in evolution (for instance, the fact that Australasian marsupials have placental counterparts on other continents) and argues that this might almost guarantee the emergence of something like us. He writes, "For all of life's plenitude there is a strong stamp of limitation, imparting not only a predictability to what we see on Earth, but by implication elsewhere."

Perhaps, more ominously, there could be a crucial hurdle at our own present evolutionary stage, the stage when intelligent life starts to develop technology. If so, the future development of life depends on whether humans survive this phase. This does not mean that Earth has to avoid a disaster, only that before this happens, some human beings or advanced artefacts will have spread beyond their home planet.

Searches for life will justifiably focus on Earth-like planets orbiting long-lived stars. But science fiction authors remind us that there are more exotic alternatives. Perhaps life can flourish even on a planet flung into the frozen darkness of interstellar space, whose main warmth comes from internal radioactivity (the process that heats Earth's core). There could be diffuse living structures, freely floating in interstellar clouds; such entities would live (and, if intelligent, think) in slow motion, but nonetheless may come into their own in the long-range future.

No life would survive on a planet whose central Sun-like star became a giant and blew off its outer layers. Such considerations remind us of the transience of inhabited worlds, and also that any seemingly artificial signal could come from superintelligent (though not necessarily conscious) computers, created by a race of alien beings that had long since died out.

Alien Intelligence: Visits or Signals?

If advanced life is widespread, we must confront the famous question first posed by the great physicist Enrico Fermi: Why haven't they visited Earth already? Why aren't they, or their artefacts, staring us in the face? This argument gains further weight when we realise that some stars are billions of years older than our Sun: if life were common, its emergence should have had a "head start" on planets around these ancient stars. The cosmologist Frank Tipler, perhaps the most vocal proponent of the view that we are alone, doesn't suggest that aliens would themselves have travelled interstellar distances. He argues, however, that at least one alien civilisation would have developed self-reproducing machines and launched them into space. These machines would spread from planet to planet, multiplying as they went; they would spread through the Galaxy within ten million years, a time far shorter than the "head start" that some of the other civilisations could have had. (Of course, there have been recurrent contentions that UFOs have indeed visited us; some people claim to have been abducted by aliens. In the 1990s, their favoured "visiting card" was a pattern of "crop circles" in cornfields, mainly in southern

England. Along with most scientists who have studied these reports, I am utterly unconvinced. Extraordinary claims need extraordinary evidence to support them, but in all these cases the evidence is flimsy. If aliens really had the brainpower and technology to reach Earth, would they merely despoil a few cornfields? Or content themselves with briefly abducting a few well-known cranks? Their manifestations are as banal and unconvincing as the messages from the dead that used to be reported in the heyday of spiritualism a hundred years ago.)

Maybe we can rule out visits by human-scale aliens, but if an extraterrestrial civilisation had mastered nanotechnology and transferred its intelligence to machines, the "invasion" might consist of a swarm of microscopic probes that could have evaded notice. Even if we haven't been visited at all, we shouldn't, despite Fermi's question, conclude that aliens don't exist. It would be far easier to send a radio or laser signal than to traverse the mind-boggling distances of interstellar space. We are already able to send signals that could be picked up by an alien civilisation; indeed, equipped with large radio antennae they could pick up the strong signals from antiballistic missile radars, as well as the combined output of all our TV transmitters.

Searches for extraterrestrial intelligence (SETI) are being spearheaded by the SETI Institute, at Mountain View, California; its work is supported by hefty donations from Paul Allen, cofounder of Microsoft, and other private benefactors. Any interested amateur with a home computer can download and analyse a short stretch of the data stream from the institute's radio telescope. Millions have taken up this offer, each inspired by the hope of being first to find "ET." In the light of this broad public interest, it seems surprising that SETI searches have had such a hard time getting public funding, even at the level of the tax revenues from a single science fiction movie. If I were an American scientist testifying before congress I would be happier requesting a few million dollars for SETI than seeking funds for more specialised science, or indeed for conventional space projects.

It makes sense to listen, rather than transmit. Any two-way exchange would take decades, so there would be time to plan a measured response. But in the long run, a dialogue could develop. The logician Hans Freudental proposed an entire language for interstellar communication, showing how it could start with the limited vocabulary needed for simple mathematical statements, and gradually build up and diversify the realm of discourse. A manifestly artificial signal, whether it was intended to be decoded or was part of some cosmic cyberspace that we were eavesdropping on, would convey the momentous message that intelligence (though not necessarily consciousness) wasn't unique to Earth.

If evolution on another planet in any way resembled the "artificial intelligence" scenarios conjectured for the twenty-first century here on Earth, the most likely and durable form of "life" may be machines whose creators had long ago been usurped or become extinct. The only type of intelligence we could detect would be one that led to a technology that we could recognise, and that could be a minor and atypical fraction of the totality of extraterrestrial intelligence. Some "brains" may package reality in a fashion that we can't conceive and have a quite different perception of reality. Others could be uncommunicative: living contemplative lives, perhaps deep under some planetary ocean, doing nothing to reveal their presence. Still other "brains" could actually be assemblages of superintelligent "social insects." There may be a lot more out there than we could ever detect. Absence of evidence wouldn't be evidence of absence.

We know too little about how life began, and how it evolves, to be able to say whether alien intelligence is likely or not. The cosmos could already be teeming with life: if so, nothing that happens on Earth would make much difference to life's long-range cosmic future. On the other hand, the emergence of intelligence may require such an improbable chain of events that it is unique to our Earth. It may simply not have occurred anywhere else, not around even one of the trillion billion other stars within range of our telescopes.

Nor can we judge how best to search for intelligent life. In earlier chapters I have emphasised that we cannot even be sure what the dominant form of intelligence on Earth will be, even a century from now. What prospect could we have of envisaging what might be spawned from another biosphere with a billion-year head start on us? We know too little to lay confident odds on what may exist or how it might manifest itself, and so we should search for anomalous radio emissions, optical flashes, and absolutely any type of signal that we have instruments to detect.

In some ways it would be disappointing if searches for alien intelligence were doomed to fail. On the other hand, such a failure would boost our cosmic self-esteem: if our tiny Earth were a unique abode of intelligence, we could view it in a less humble perspective than it would merit if the Galaxy already teemed with complex life.

13

BEYOND EARTH

If robotic probes and fabricators spread through the solar system, would any humans follow? Communities away from Earth would be established (if at all) by risk-taking individualist pioneers. Travel beyond the solar system is a far remoter, posthuman, prospect.

AN ICONIC IMAGE FROM THE 1960S was the first photograph from space, showing our spherical Earth. Jonathan Schell suggests that this picture should be complemented by another one, which focuses on our planet but is extended in time rather than in space: "The view that counts is the one from Earth, from within life. . . . From this Earthly vantage point another view—one even longer than the one from space—opens up. It is the view of our children and grandchildren, and of all the future generations of humankind, stretching ahead of us in time. . . . The thought of cutting off life's flow, of amputating this future, is so shocking, so alien to nature, and so contradictory to life's impulse that we can scarcely entertain it before turning away in revulsion and disbelief."

Is it worth taking precautions to ensure that whatever may happen, something survives of humanity? Most of us care about the future, not just because of a personal concern with children and grandchildren, but because all our efforts would be devalued if they were not part of a continuing process, if they did not have consequences that resonated into the far future.

It would be absurd to claim that emigration into space is an answer to the population problem, or that more than a tiny fraction of those on Earth will themselves ever leave it. If some disaster reduced humanity to a far lower population, living in primitive conditions in a devastated wasteland, the survivors would still find Earth's environment more hospitable than that of any other planet. Nonetheless, even a few pioneering groups, living independently of Earth, would offer a safeguard against the worst possible disaster—the foreclosure of intelligent life's future through the extinction of all humankind.

The ever-present slight risk of a global catastrophe with a "natural" cause will be greatly augmented by the risks stemming from twenty-first-century technology. Humankind will remain vulnerable so long as it stays confined here on Earth. Is it worth, in the spirit of Pascal's wager, insuring against not just natural disasters but the probably much larger (and certainly growing) risk of the human-induced catastrophes discussed in earlier chapters? Once self-sustaining communities exist away from Earth—on the Moon, on Mars, or freely floating in space—our species would be invulnerable to even the worst global disasters.

So how feasible would it be to establish a sustainable habitat elsewhere in the solar system? How long will it be before people return to the Moon, and perhaps explore still further afield?

Will Manned Space Flight Revive?

Those of us who are now middle-aged can remember the murky live TV pictures of Neil Armstrong's "one small step." In the 1960s, President Kennedy's programme to "land a man on the Moon before the end of the decade, and return him safely to Earth" took space flight from the corn-flakes packet to reality. And it seemed just a beginning. We imagined follow-up projects: a permanent "lunar base," rather like the existing base at the South Pole; or even huge "space hotels" orbiting Earth. Manned expeditions to Mars seemed a natural next step. But none of these has happened. The year 2001 didn't resemble Arthur C. Clarke's depiction, any more than 1984 (fortunately) resembled Orwell's.

Rather than being a precursor for a continuing and ever more ambitious programme of manned space flight, the Apollo moon landing programme was a transient episode, motivated primarily by the urge to "beat the Russians."

The last lunar landing was in 1972. Nobody much under the age of thirty-five can remember when men walked on the Moon. To young people, the Apollo programme is a remote historical episode: they know the Americans landed men on the Moon, just as they know the Egyptians built the pyramids; but the motivations seem almost as bizarre in the one case as in the other. The 1995 film *Apollo 13*, a "docudrama" starring Tom Hanks, about the near disaster that befell James Lovell and his crew on a voyage round the Moon, was for me (and I suspect for many others of similar vintage) an evocative reminder of an episode we had followed anxiously at the time. But to a young audience, the outdated gadgetry and the traditional "right stuff" values seemed almost as antiquated as a traditional "Western."

The practical case for manned space flight was never strong, and it gets ever weaker with each advance in robotics and miniaturisation. The use of space for communications, meteorology, and navigation has forged ahead, benefiting from the same technical advances that have given us mobile phones and high-performance laptop computers here on Earth. Space exploration for scientific purposes can be better (and far more cheaply) carried out by unmanned probes. Huge numbers of miniaturised robotic probes—"intelligent machines"—will, twenty-five years from now, be dispersed through the solar system, sending back images of planets, moons, comets, and asteroids, revealing what they are made of, and perhaps constructing artefacts from the raw materials to be found in them. There may be long-term economic benefits from space, but these will be implemented by robotic fabricators, not by people.

But what is the future for manned space flight? In the 1990s Russian cosmonauts spent months, even years, circling Earth in the increasingly decrepit *Mir* space station. Having far surpassed its design lifetime, *Mir* ended its mission in 2001 with a final splashdown in the Pacific Ocean. Its successor, the *International Space Station* (ISS) will be the most expensive artefact ever constructed, but it is a "turkey" in the sky. Even if it is finished, something that seems uncertain, given the immense and ever-rising costs, and prolonged delays, it can do nothing to justify its price tag. Thirty years after men walked on the Moon, a new generation of astronauts is going round and round Earth, in more comfort than *Mir* offered, but much more expensively. At the time of writing, the number of astronauts on board has been scaled back to three, for reasons of safety and finance: they will be preoccupied with "housekeeping" tasks, making it even less likely that anyone on board will pursue any serious or interesting projects. Indeed, it is as sub-optimal to do most science from the ISS as it would be to do ground-based astronomy from a boat. Even in the US, the scientific community was firmly opposed to the ISS, and abandoned campaigning against it only when the political momentum became unstoppable. It is sad that they weren't listened to: it is a wasteful political failure that government funds couldn't have been channelled towards the same aerospace companies for alternative projects that were either useful or inspirational. The ISS is neither.

There is only one reason to applaud the ISS: if one believes that in the long run space travel will become routine, this continuing programme ensures that the forty years of experience of manned space flight gained by US and Russia is not dissipated.

A revival in manned space flight must await changes in technology and—perhaps even more—changes in style. Present launching techniques are as extravagant as air travel would be if the plane had to be rebuilt after every flight. Space flight will become affordable only when its technology comes closer to that of supersonic aircraft. Tourist trips into orbit may then become routine. Already, the American financier Dennis Tito and the South African software magnate Mark Shuttleworth have spent twenty million dollars in return for a week in the ISS. There is a line-up of others willing to follow these "space tourists," even at that price; there would be far more if the tickets got cheaper.

Indeed, private individuals won't, in the long run, restrict themselves to the role of passengers passively circling Earth. When this kind of escapade palls, seeming too tame and routine, some will yearn to go further. Manned expeditions into deep space could be entirely funded by private individuals or consortia, perhaps indeed becoming the province of wealthy adventurers prepared, like test pilots or Antarctic explorers, to accept high risks to boldly explore the far frontier and experience thrills beyond those provided by large yachts or round the world ballooning. The Apollo programme was a government-funded quasi-military enterprise; future expeditions could he quite different in style. If high-tech billionaires like Bill Gates or Larry Ellison seek challenges that won't make their later life seem an anticlimax, they could sponsor the first lunar base or even an expedition to Mars.

The "Cheap" Route to Mars

If Martian exploration were initiated in the near future, it might well follow the format advocated by the maverick American engineer Robert Zubrin. In response to discouraging claims from NASA that an expedition would cost over one hundred billion dollars, Zubrin proposed a cut-price "Mars direct" strategy that would bypass the *International Space Station*. He aimed to evade one of the main problems of earlier schemes: the need to carry, on the outward journey, all the fuel for a return trip. The proposal, presented in his book *The Case for Mars*, involves first sending directly to Mars an unmanned probe that will manufacture the fuel for the return journey. It would carry a chemical processing plant, a small nuclear reactor, and a rocket capable of bringing back the first group of explorers. This rocket would not be fully fuelled: its tanks would be filled with pure hydrogen. The nuclear reactor (pulled by a small tractor that would also be part of the first payload) would then generate energy for the chemical plant, which would use hydrogen to convert carbon dioxide from the Martian atmosphere into methane and water. The water would then be broken down, the oxygen stored, and the hydrogen recycled to make more methane. The return rocket fuel would then be methane and oxygen. Six tonnes of hydrogen would allow one hundred tonnes of methane to be made, enough to fuel the astronauts' return rocket. (Of course, if water could be extracted from permafrost that was not too deep below the surface, part of this process could be bypassed.)

Two years later, a second and third spacecraft would be launched. One would carry a cargo similar to that of the earlier craft, while the other would contain the crew, along with sufficient provisions for a sojourn on Mars of up to two years. The manned craft would go on a faster trajectory than that carrying the cargo. This means that the crew need not be launched until (and unless) the cargo was safely on its way, but they could nonetheless reach Mars before the cargo arrived. If through some mishap they landed far away from the intended site (where the first instalment of cargo was located), there would still be time to divert the second cargo craft to the actual landing site, so that wherever they landed, the crew would have supplies. Once this path-finding mission was accomplished, there could be one or more follow-ups every two years, gradually building up infrastructure.

Would anyone want to go? There may be a parallel here with terrestrial exploration, which was driven by a variety of motives. The explorers who set out from Europe in the fifteenth and sixteenth century were bankrolled mainly by monarchs, in the hope of recouping exotic merchandise or colonising new territory. Some, for instance Captain Cook on his three eighteenth-century expeditions to the South Seas, were publicly funded, at least in part as a scientific enterprise. And for some early explorers—generally the most foolhardy of all—the enterprise was primarily a challenge and adventure: the motivation of present-day mountaineers and round-the-world sailors.

The first travellers to Mars, or the first long-term denizens of a lunar base, could be impelled by any of these motives. The risks would be high; but in fact, no space travellers would be venturing into the unknown to the extent that the great terrestrial navigators were. These early ocean voyagers had far less foreknowledge of what they might encounter, and many died in the enterprise. Nor would any space travellers be cut off from human contact. There would admittedly be a thirty-minute turnaround for messages to and from Mars. But it took months for traditional explorers to send messages home; and some—Captain Scott and other polar pioneers among them—had no such contact at all.

The stakes are high in opening up new worlds. It seems taken as an axiom that all should return. But maybe the most determined pioneers would be prepared to accept—as many Europeans willingly did when they set out for the New World—that there would be no return. Many could be found who would sacrifice themselves in a glorious and historic cause; by forgoing the option of ever returning home, they would slash the cost by obviating the need to carry rocket casings and hydrogen for the return trip. A Martian base would develop more quickly if those constructing it were content with one-way tickets.

Futurists and space enthusiasts often urge that "humanity" or "the nation" should choose to do something. Space exploration indeed began as a quasi-military enterprise funded by governments. But this rhetoric is inappropriate to manned space exploits in the twenty-first century. Most great innovations and achievements were initiated not because they were a national goal, still less a goal of humanity, but because of economic motivation or simply personal obsession.

The enterprise will become far cheaper and less precarious when propulsion systems are more efficient. It currently takes several tonnes of chemical fuel to propel one tonne of payload away from the grip of Earth's gravity. Space travel is difficult primarily because the trajectory has to be planned with high precision in order to minimise fuel consumption. But if there were, say, ten times more thrust for each kilogram of fuel, then midcourse adjustments could be made whenever necessary, just as we do when driving along a winding road. Keeping a car on the road would be a high-precision enterprise if the journey had to be programmed beforehand, with no chance of adjustments on the way. If one could be profligate with power and fuel, space travel would be an almost unskilled exercise. The destination (the Moon, Mars, or an asteroid) is in clear view. One just has to steer towards it and use retrojets to brake by the right amount at journey's end.

We don't yet know what kind of novel propulsion systems will prove most promising: solar and nuclear power are the two obvious near-term options. It would greatly help if the propulsion system and the fuel needed for escape from Earth's gravity could be located on the ground rather than having to be part of the cargo. One possibility is immensely powerful ground-based lasers. Another is a space elevator, a wire made of carbon fibre extending more than thirty-five thousand kilometres up into space and held aloft by a geostationary satellite. (Carbon nanotubes have a tensile strength that is high enough. Very thin carbon "yarns" have already been made that are up to thirty centimetres long; the challenge is to fabricate tubes of enormous length, or devise techniques for weaving many into a very long wire that retains the strength of separate fibres.) This "elevator" would allow payloads and passengers to be hoisted from the grip of Earth's gravity by power supplied from the ground. The rest of the voyage could be powered by a low-thrust (perhaps nuclear) rocket.

Before human beings venture into deep space, the entire solar system will have been mapped and probed by flotillas of tiny robotic craft, controlled by the ever more powerful and miniaturised "processors" that nanotechnology will provide. A manned expedition to Mars will have been preceded by the cargoes of provisions envisaged by Zubrin, and perhaps also by seeds of plants designed to thrive and multiply on the red planet. Freeman Dyson envisages genetically engineered "designer trees" that could grow a transparent membrane around themselves that functions as a greenhouse.

Brute-force methods have been proposed for "terraforming" the entire surface of Mars to render it more habitable. It could be warmed by injecting greenhouse gases into its tenuous atmosphere, or placing huge mirrors in orbit to direct more sunlight to the poles, or even covering tracts of the Martian surface with something black to absorb sunlight—soot or powdered basalt. Terraforming would take centuries; but within a century there could be a permanent presence on localised bases. Once the infrastructure was there, two-way trips would become less costly and could be more frequent.

Issues of environmental ethics may loom large. Would it be acceptable to exploit Mars, as happened when (with tragic consequences for the Native Americans) the pioneer settlers advanced westward across the United States? Or should it be preserved as a natural wilderness, like the Antarctic? The answer should I think depend on what the pristine state of Mars actually is. If there were any life there already—especially if it had different DNA, testifying to quite separate origin from any life on Earth—then there would be widely voiced views that it should be preserved as unpolluted as possible. What might actually happen would depend on the character of the first expeditions. If they were governmental (or international), Antarctic-style restraint might be feasible. On the other hand, if the explorers were privately funded adventurers of a free-enterprise (even anarchic) disposition, the Wild West model would, whether we liked it or not, be more likely to prevail.

Deeper into Space

The focus will not stay exclusively on the Moon and Mars. Life
could eventually spread and diversify among comets and asteroids, even in the cold outer reaches of the solar system: the vast
number of small bodies in the solar system have, in toto, a far
larger habitable surface than the planets.

An alternative would be to construct an artificial habitat,
freely floating in space. This option was studied back in the
1970s by Gerard O'Neill, an engineering professor at Princeton University. He envisaged a spacecraft in the shape of a vast
cylinder, slowly spinning around its axis. The occupants would
live on the inside of its walls, pinned to them by the artificial
gravity generated by its spin. The cylinders would be big
enough to have an atmosphere, even perhaps clouds and rain,
and could accommodate tens of thousands in an environment
that, in O'Neill's perhaps fanciful sketches, resembled a leafy
Californian suburb. The material to build these gargantuan
structures would have to be "mined" from the Moon or from
asteroids. O'Neill made the valid point that once large-scale
robotic engineering projects can be carried out in space, using
raw materials that need not be lifted from Earth, it becomes
feasible to build artificial space platforms on a very ample scale.

O'Neill's specific scenarios may become technically feasible,
but they will remain sociologically implausible. A single fragile
structure containing tens of thousands of people would be even
more vulnerable than integrated communities down on Earth
to a single act of sabotage. A more dispersed set of smaller-
scale habitats would offer more robust chances of survival and
development.

In the second half of the twenty-first century there could be
hundreds of people in lunar bases, just as there now are at the
South Pole; some pioneers could already be living on Mars, or

else on small artificial habitats cruising the solar system, attaching themselves to asteroids or comets. Space will also be pervaded by robots and intelligent "fabricators," using raw material mined from asteroids to construct structures of ever-expanding scale. I am not especially advocating these developments, but they nonetheless seem plausible, both technically and sociologically.

The Far Future

Still further ahead, in future centuries, robots and fabricators could have pervaded the entire solar system. Whether human beings will themselves have joined this diaspora is harder to predict. If they did, communities would develop in a manner that eventually made them quite independent of Earth. Unconstrained by any restrictions, some would surely exploit the full range of genetic techniques and diverge into new species. (The constraint due to lack of genetic diversity in small groups could be overcome by artificially induced variations in the genome.) The diverse physical conditions—very different on Mars, in the asteroid belt, and in the still colder far reaches of the solar system—would give renewed impetus to biological diversification.

Although a contrary view is often expressed, the expanses of space offer little prospect of a solution to resource or population problems on Earth: these will have to be sorted out down here, if the problem isn't rendered nugatory by one of the disastrous setbacks to terrestrial civilisation conjectured in earlier chapters. The populations in space may eventually grow exponentially, but this will be because of their autonomous growth rather than by "emigration" from Earth. Those going into space will be impelled by an exploratory urge. But their choices will have epochal consequences. Once the threshold is crossed when there is a self-sustaining level of life in space, then life's long-range future will be secure irrespective of any of the risks on Earth (with the single exception of the catastrophic destruction of space itself). Will this happen before our civilisation disintegrates, leaving the prospect as a might-have-been? Will self-sustaining space communities be established before a catastrophe sets back the prospect of any such enterprise, perhaps foreclosing it for ever? We live at what could be a defining moment for the cosmos, not just for our Earth.

The beings that could, within a few hundred years, occupy sites in our solar system would all be recognisably humanoid, though they would be complemented (and probably, in the most inhospitable locations, vastly outnumbered) by robots with human intelligence. However, travel beyond the solar system, through interstellar space, would, if it ever happened, be a posthuman challenge. Voyages would initially involve robotic probes. The journey would last many human generations and require a self-contained community, or suspended animation of any living intelligence. Alternatively, genetic material, or blueprints downloaded into inorganic memories, could be launched into the cosmos in miniature spacecraft. They could be programmed to land on promising planets, and duplicate copies of themselves, thereby starting a diffusion through the entire Galaxy. There could even be laser transmission of "encoded" information (a kind of "space travel" that could happen at the speed of light) which could trigger the assembly of artefacts or the "seeding" of living organisms in propitious locations. Such concepts confront us with profound issues about the limits of information storage, and philosophical implications of identity.

This would be as epochal an evolutionary transition as that which led to land-based life on Earth. But it could still be just the beginning of cosmic evolution.

A Gigayear Perspective

A hackneyed anecdote among astronomy lecturers describes a worried questioner asking: "how long did you say it would be before the sun burnt the Earth to a crisp?" On receiving the answer, "six billion years," the questioner responds with relief: "thank God for that, I thought you said six million." What happens in far-future aeons may seem blazingly irrelevant to the practicalities of our lives. But I don't think the cosmic context is entirely irrelevant to the way we perceive our Earth and the fate of humans.

The great biologist Christian de Duve envisages that "The tree of life may reach twice its present height. This could happen through further growth of the human twig, but it does not have to. There is plenty of time for other twigs to bud and grow, eventually reaching a level much higher than the one we occupy while the human twig withers. . . . What will happen depends to some extent on us, since we have now have the power of decisively influencing the future of life and humankind on Earth."

Darwin himself noted that "not one living species will transmit its unaltered likeness to a distant futurity". Our own species may change and diversify faster than any predecessor, via intelligently controlled modifications, not by natural selection alone. Long before the Sun finally licks Earth's face clean, a teeming variety of life or its artifacts could have spread far beyond its original planet; provided that we avoid irreversible catastrophe before this process can even commence. They could look forward to a near-infinite future. Wormholes, extra dimensions and quantum computers open up speculative scenarios that could transform our entire universe eventually into a "living cosmos."

The first aquatic creatures crawled onto dry land in the Silurian era, more than three hundred million years ago. They may have been unprepossessing brutes, but had they been clobbered, the evolution of land-based fauna would have been jeopardised. Likewise, the post-human potential is so immense that not even the most misanthropic amongst us would countenance its being foreclosed by human actions.

14

EPILOGUE

TRADITIONAL WESTERN CULTURE envisaged a beginning and an end of history, but a constricted timespan—just a few thousand years—in between. (Many, however, queried the exactitude of the Archbishop of Armagh, James Ussher, who famously dated the creation at Saturday afternoon on 22nd October, 4004 B.C.E.) Moreover, history was widely believed to have already entered its final millennium. For the 17th century essayist Sir Thomas Browne "the world itself seems in the wane. A greater part of Time is spun than is to come."

To Ussher's mind, the creation of the world and the creation of humanity were within a week of one another; to our modern minds, the two events are unimaginably far apart. There was a vast absence before us, and its record stares out at us from every rock. The evolution of Earth's biosphere can now be traced back several billion years: the future of our physical universe is reckoned to be more extended still, perhaps even infinite. But despite these expanded horizons, both past and future, one timescale has contracted: pessimistic estimates of how long our civilization has to run, before crumbling, or even undergoing a terminal apocalypse, are shorter than would have been gauged by our forebears who devotedly added bricks to cathedrals that would not be finished in their lifetime. Earth itself may endure, but it will not be humans who cope with the scorching of our planet by the dying sun; nor even, perhaps, with the exhaustion of Earth's resources.

If our solar system's entire lifecycle, from its birth in a cosmic cloud to its death-throes in the Sun's terminal flare-up, were to be viewed "fast forward" in a single year, then all recorded history would be less than a minute in early June. The twentieth century would flash past in a third of a second. The next fraction of a second, in this depiction, will be "critical": in the twenty-first century, humanity is more at risk than ever before from misapplication of science. And the environmental pressures induced by collective human actions could trigger catastrophes more threatening than any natural hazards.

For several recent decades, we were vulnerable to a nuclear holocaust. We escaped, but in retrospect our survival seems due as much to good luck as to intrinsically favourable odds. Moreover, recent knowledge (especially in biology) has opened up non-nuclear dangers that could be even more sombre in the next half-century. Nuclear weapons give an attacking nation a devastating advantage over any feasible defense. New sciences will soon empower small groups, even individuals, with similar leverage over society. Our increasingly interconnected world is vulnerable to new risks; "bio" or "cyber," terror or error. These risks can't be eliminated: indeed it will be hard to stop them from growing without encroaching on some cherished personal freedoms.

The benefits opened up by biotechnology are manifest, but they must be balanced against the accompanying hazards and ethical constraints. Robotics or nanotechnology will also involve trade-offs: they could have disastrous or even uncontrollable consequences when misapplied. Experimenters should be cautious in"pushing the envelope" of science; even if there were a case for putting the brakes on some research, a moratorium could never be effectively enforced worldwide.

Neither speculative thinkers like H.G. Wells nor his scientific contemporaries had much success in foreseeing the highlights of twentieth-century science.The present century is even less predictable because of the possibility of altering or supplementing the human intellect. But any entirely unsuspected new advances may well pose novel hazards too. Special responsibility lies with scientists themselves: they should be mindful of how their work might be applied, and do all they can to alert the wider public to potential perils.

A key challenge is to understand the nature of life; how it began, and whether it exists beyond Earth. (There is certainly no other scientific question that I would personally be more eager to see answered). Alien life may be discovered—even, conceivably, alien intelligence. Our planet could be one of millions that are inhabited: we may live in a biofriendly universe already teeming with life. If so, the most epochal happenings on Earth, even our utter extinction,would barely register as a cosmic event. In the quaint words of the eighteenth-century astronomer and mystic,Thomas Wright of Durham:"In this great Celestial Creation, the Catastrophy of a World, such as ours, or even the total Dissolution of a System of Worlds, may possibly be no more to the great Author of Nature, than the most common Accident in Life with us, and in all Probability such final and general Dooms Days may be as frequent there, as even Birth-Days or Mortality with us upon this Earth."

But it could turn out that the odds are heavily stacked against the emergence of life, so that our biosphere is the unique abode of intelligent and self-aware life within our Galaxy. Our small Earth's fate would then have a significance that was truly cosmic—an importance that would reverberate through the whole of Thomas Wright's "Celestial Creation."

Our primary concerns are naturally with the fate of our present generation, and to reduce the threats to us. But for me, and perhaps for others (especially those without religious belief), a cosmic perspective strengthens the imperative to cherish this "pale blue dot" in the cosmos. It should also motivate a circumspect attitude towards technical innovations that pose even a small threat of a catastrophic downside.

The theme of this book is that humanity is more at risk than at any earlier phase in its history. The wider cosmos has a potential future that could even be infinite. But will these vast expanses of time be filled with life, or as empty as the Earth's first sterile seas? The choice may depend on us, this century.

NOTES

Chapter 1

3 "... real life Dr. Strangeloves ..." The most prominent nuclear strategist was Herman Kahn, author of *On Thermonuclear War* (Princeton University Press, 1960).

4 "... words of Gregory Benford ..." *Deep Time*, by Gregory Benford, is published by Avon Books, NY, (1999).

5 "... mathematician and philosopher Frank Ramsey ..." *Foundations of Mathematics and other Logical Essays*, by F. P. Ramsey, London, Kegan Paul, Trench and Trubner, p. 291. Published posthumously in 1931.

6 "About 4.5 billion years ago ..." For a fuller account of cosmic history, see my book *Our Cosmic Habitat* (Princeton University Press and Phoenix paperback, 2003).

Chapter 2

9 "... 1902 lecture entitled 'Discovery of the Future' ..." H.G. Wells's Royal Institution lecture, given on 24 January 1902, was, unusually, reprinted in full in the journal *Nature*. The programme note described him as "H.G. Wells, B Sc": he was inordinately proud of the academic degree he had gained by external study at London University.

12 "... Lee Silver, in his book ..." *Remaking Eden*, by Lee Silver, Avon Books, New York (1997).

13 "In 1937 the US National Academy of Sciences ..." The NAS Study is described, and interestingly critiqued, by C.H. Townes, co-inventor of the maser, in his book *Making Waves*, Springer-Verlag 1995.

15 "... there can still be revolutionary innovations, ..." Science and technology now have a complex symbiosis, which didn't exist a hundred years ago: research triggers applications; equally, new techniques and instruments boost scientific discovery.

16 "Ray Kurzweil ..." *The Age of Spiritual Machines*, by Ray Kurzweil, Viking, NY (1999).

16 "... circuits on a much finer scale, ..." One promising technique, proposed by electrical engineers at Princeton University, involves engraving the required pattern onto a sliver of quartz, putting a layer of silicon on top of it, and then firing a laser to melt the part of the silicon in contact with the quartz mould.

17 "the evangelists of nanotechnology ..." A recent survey of the nearer-term prospects for nanotechnology is *Our Molecular Future*, by Douglas Mulhall, Prometheus Books (2002).

17 "The robotics pioneer Hans Moravec ..." *Mind Children: The Future of Robot and Human Intelligence*, by Hans Moravec, Harvard University Press (1988).

18 "How much non-biological hardware ..." John Sulston, in *Big Questions in Science*, H. Swain, ed., Jonathan Cape, London (2002), pp.159–163.

19 "The Californian futurologist Vernon Vinge ..." Vernon Vinge's article on the singularity appeared in *Whole Earth* magazine (1993).

22 "Futurists like Freeman Dyson ..." *The Sun, the Genome and the Internet*, by Freeman Dyson, Oxford University Press (1999).

23 "... discusses how to optimise ..." *The Clock of the Long Now*, by Steward Brand, BasicBooks, NY, Orion Books, London (1999).

24 "... North America reverts to a medieval state ..." *A Canticle for Leibowitz*, by Walter M. Miller Jr., Orbit paperback, 1993 (first published in 1960).

24 "the 'Gaia' concept, ..." James Lovelock is quoted by Stewart
Brand in *The Clock of the Long Now*.

Chapter 3

25 "... 187 million perished ..." The estimate comes from Z.
Brzezinski, *Out of Control: Global Turmoil on the Eve of the Twenty-
First Century*, New York, 1993; this same number is endorsed by
Eric Hobsbaum in his *Age of Extremes*, Michael Joseph, London
(1994).
26 "This was not only the most dangerous moment ..." Arthur M.
Schlesinger Jr.'s remarks were quoted in the *New York Times*, October 12, 2002, reporting on a conference held to mark (and
reminisce about) the Cuban Missile Crisis on its fortieth an-
niversary. At this conference new facts emerged that showed that
the world was even closer to the "knife edge" than the public had
previously realised. During the crisis a Russian submarine was
targeted by depth charges from a US warship. This submarine
carried a nuclear-armed torpedo, which could have been
launched with the concurrence of three officers. Fortunately, one
young officer, Vasily Arkhipov, held out against the pressure to
launch the torpedo, thereby staving off an escalation that could
well have run out of control.
26 "Even a low probability ..." Robert McNamara was interviewed
by Jonathan Schell in the *Nation*.
28 "The *Bulletin of Atomic Scientists* was founded ..." The *Bulletin of
Atomic Scientists* is now published bimonthly by the educational
Foundation for Nuclear Science in Chicago (http://www.
thebulletin.org).
29 "... virtually every technical innovation ..." McNamara is quoted
by Solly Zuckerman in *Nuclear Illusions and Reality*, Collins, London (1982).
30 "... a world-wide blocking-out ..." The concept of nuclear winter was proposed in a 1983 study by R.P. Turco, O.B. Toon, T.P.
Ackerman, J.B. Pollack, and C. Sagan (known as TTAPS). The
quantitative details of this study, depending on the amount of

smoke and soot released, and how long it would stay in the atmosphere, were the subject of subsequent controversy.

32 "The basic reason . . ." *Nuclear Illusion and Reality*, by Solly Zuckerman. The quotations are taken from pages 103 and 107.

34 "The total of excess material . . ." *Megatons and Megawatts*, by R. L. Garwin and G. Charpak, Random House (2002).

36 ". . . argues that undetectable tests . . ." *Technical Issues Related to the Comprehensive Nuclear Test Ban Treaty*, a report of the Committee on International Security and Arms Control, National Academy of Sciences, published in 2002.

37 ". . . founded a series of conferences. . . ." Information on the Pugwash conferences and their history is accessible on http://www.pugwash.org/. The obscure village after which the conferences were named had incongruous associations in the UK, where "Captain Pugwash" was a well-known cartoon character on children's television.

38 "cease and desist . . ." The quotation is from an article by Hans Bethe in the *New York Review of Books*.

38 ". . . manifesto stressing the urgency . . ." The Einstein–Russell manifesto was recently reprinted, with commentary, by the Pugwash organisation.

39 "an international group convened . . ." The Canberra Commission on the Elimination of Nuclear Weapons presented its report to the Australian government in 1997. In addition to those mentioned in the text, its members included General Lee Butler, former head of US Strategic Air Command, and an eminent British soldier, Field Marshal Carver.

40 ". . . two most prominent personalities. . . ." For a detailed account see *Brotherhood of the Bomb: The Tangled Lives and Loyalties of Robert Oppenheimer, Ernest Lawrence and Edward Teller*, by Gregg Herken, Henry Holt (2002).

Chapter 4

43 ". . . portrayed devastation . . ." Tom Clancy is notable for the pre-

science and technical fidelity of his plot lines. An earlier novel, *Debt of Honor*, featured the use of an airliner as a missile to attack the Capitol building in Washington.

44 "With modern weapons-grade uranium . . ." Luis Alvarez is quoted on the website of the Nuclear Control Institute, Washington, DC.

44 "A nuclear explosion at the World Trade Center . . ." This scenario, along with other related material, is discussed in *Avoiding Nuclear Anarchy*, edited by G.T. Allison (BCSIA Studies in International Security, 1996).

44 "We have slain the dragon . . ." James Wolsey was speaking at US Senate hearings in February 1993.

45 ". . . crash onto the containment vessel." A brief survey of these risks (with references) is *Nuclear Power Plants and Their Fuel as Terrorist Targets*, by D.M. Chaplin and eighteen co-authors, in *Science* 297, pp. 997–998, 2002. In a later response, Richard Garwin claimed that the authors were downplaying the risks, which had been treated more seriously in a National Academy of Sciences report. [See also *Science* 299, pp. 201–203, (2003).]

48 ". . . he was in charge of . . ." *Biohazard*, by Ken Alibek, with Stephen Handelman, Random House, New York, (1999).

48 "The knowledge and techniques . . ." Fred Ikle, March 1997, cited in *The Shield of Achilles*, by Philip Bobbitt, Penguin, NY and London (2002).

50 ". . . several scenarios were explored . . ." The Jason study of biological threat was summarised in an article by Steven Koonin, provost of the California Institute of Technology and chairman of the Jason group, in *Engineering and Science*. 64[3–4] (2001).

51 ". . . simulated a covert smallpox attack . . ." The Dark Winter exercise was carried out by the Johns Hopkins Center for Civilian Biodefense Strategies, in collaboration with the Center for Strategic and International Studies (CSIS), the Analytic Services (ANSER) Institute for Homeland Security, and the Oklahoma National Memorial Institute for the Prevention of Terrorism.

54 "According to a report." The report is *Making the Nation Safer:*

The Role of Science and Technology in Countering Terrorism, National Academy Press (2002).

54 "It would be interesting . . ." George Poste, in *Prospect* (May 2002).

55 ". . . had assembled a polio virus, . . . " J. Cello, A.V. Paul, and E. Wimmer, *Science* 207, p. 1016 (2002).

56 ". . . more effectively than natural mutations." A technique used by a US company called Morphotek involves increasing the mutation rate by inserting into animals, plants, or bacteria a gene called PMS2-134, a defective version of a gene responsible for repairing DNA.

56 ". . . sequenced the human genome . . ." Craig Venter's project has been widely reported, for instance by Clive Cookson in *Financial Times*, September 30, 2002.

57 ". . . risks stemming from error . . ." The paper reporting the experiments by Ron Jackson and Ian Ramshaw was published in the *Journal of Virology* (February 2001).

58 ". . . even animals that had been previously vaccinated." In *The Demon in the Freezer* (Random House, 2002) Richard Preston reports experiments by Mark Buller and colleagues at the St. Louis School of Medicine that attempted to reproduce the Australian results. They got concurrent results, except that some mice that had been recently vaccinated retained their immunity against the modified mousepox virus.

58 "grey goo scenario" *Engines of Creation*, by Eric Drexler, Anchor Books, NY (1986).

59 ". . . biovorous replicators." There are some limits on the virulence and speed of a takeover, but these are very crude and far from reassuring. Robert A. Freitas, in a paper entitled "Some limits to global ecophagy by biovorous nanoreplicators," concludes that the replication time could be as short as one hundred seconds.

59 ". . . self-destruct 'naturally.'" Another riposte is that an organism that succeeds in evolutionary terms plainly must not despoil its habitat completely, but must instead maintain a symbiosis with it.

Chapter 5

63 "... designing web pages ..." The content of the now-defunct Heaven's Gate cult website is now archived at http://www.wave. net/upg/gate/heavensgate.html.

64 "... cyber-community of the likeminded." *republic.com*, by Cass Susstein, published by Princeton University Press in 2001.

65 "... Millenarian believers." A series of books depicting the apocalyptic era—the *Left Behind* series—has topped bestseller lists in the US.

66 "... surveillance might be the least ..." *The Transparent Society*, by David Brin, Addison-Wesley, NY (1998).

67 "... through films and TV news reports." According to a survey by the *Economist* (December 20/27 2002 issue), more than two billion people in the developing world have access to satellite television. Although locally produced programmes are increasingly favoured, the most popular Western programme in several countries (including, for instance, Iran) is "Baywatch."

68 "... mood-altering medications ..." *Our Posthuman Future*, by Francis Fukuyama, Farrar, Strauss and Giroux (New York) and Profile Books, London (2002).

69 "If we can alter people's desire ..." Steve Bloom's article is in the October 10, 2002, issue of *New Scientist*.

70 "... form of mind control ..." *Beyond Freedom and Dignity*, by B.F. Skinner Bantam/Vintage (1971).

70 "... mentally abnormal human beings ..." Philip K. Dick. The story for *Minority Report* is in his collection of short stories.

71 "ever more tightly linked ..." *Clock of the Long Now*, by Stewart Brand, BasicBooks, NY, Orion Books, London (1999).

Chapter 6

73 "... a series of 'long bets.'" The long bets were published in the May 2002 issue of *Wired*.

73 "... whether such a theory even exists." See further discussion of fundamental theories in Chapter 11.

74 "... to witness the outcome." Steven Austad and Jay Olshansky have a bet on this subject with a stake such that the heirs of the winner may, in 2150, receive as much as five hundred million dollars.

74 "... introduction of new drugs ..." "The Hidden Cost of Saying No," by Freeman Dyson appeared in the *Bulletin of the Atomic Scientists*, July 1975, and is reprinted in *Imagined Worlds*, Penguin (1985).

75 "... research just the way it led me." *The Island of Dr. Moreau*, by H.G. Wells, first published in 1896.

75 "declaration put forward in 1975 ..." Asilomar declaration. See discussion in H.F. Jutson's *The Eighth Day of Creation* (1979).

76 "... attempt at self-regulation ..." The retrospective views of several Asilomar participants are reported in "Reconsidering Asilomar," *The Scientist* 14[7]:15 (April 3, 2000).

78 "The views of scientists should not have special weight...." There are two issues, however, where specialists should be heeded: First, they are best placed to judge whether or not a problem is soluble. Some problems, though plainly important, are not yet ripe for a frontal attack, so it is no good throwing money at them. President Nixon's initiative for a "war on cancer" was premature. Untargeted fundamental research was a better bet at that time. Second, when scientists argue that undirected "blue skies" research can be the most productive, this is not just because they prefer to be free to follow where curiosity leads them. Even from a hard-nosed practical perspective this can be true: thirty years after Nixon's programme, a main challenge in cancer research is still the basic one of understanding cell division at the molecular level.

79 "... they are now used in digital cameras." There has been an interesting shift between the 1970s and today. The cutting-edge instruments used to be developed by the military, and were then adapted for scientific use. Now, the mass market for consumer electronics (digital cameras, computer-game software, and consoles) often sets the state of the art.

81 "... clone his elderly dog." The donor, John Sparling, founder of

the University of Phoenix didn't get his replacement dog, though the research group cloned a cat for the first time in March 2002.

83 "... an open attitude with regard to students." This openness plainly should not extend to those who had no intention of gaining education, but might masquerade as students simply to gain access to pathogens in university laboratories.

84 "claimed to have generated nuclear ..." The cold fusion episode is recounted in *Too Hot to Handle*, by Frank Close, Princeton University Press 1991.

84 "... investigating a puzzling effect ..." The Taleyarkhan paper is in *Science* 295, 1868 (2002).

85 "... no impediment to openness." Openness would not guarantee wide and effective scrutiny if the scientific evidence came from some enormous (and perhaps unique) facility, for instance, a spacecraft or a huge particle accelerator. In such cases the main safeguard has to come from internal quality control within the research group, which is likely to be large and intellectually diverse in these cases.

86 "in favour of 'going slow.'" Bill Joy, "Why the Future doesn't need us," was the cover article in the April 2000 issue of *Wired*.

Chapter 7

89 "... named Shoemaker–Levy after its discoverers." The comet was discovered by Eugene Shoemaker, an expert on lunar and planetary studies; his wife, Carolyn; and David Levy, an astronomer based in Arizona. In 1993 the comet passed close to Jupiter, and the tidal effect of the planet's gravity tore it apart, into about twenty pieces. It was possible to calculate that the fragments would actually crash into Jupiter sixteen months later.

93 "It is a minor risk, ..." *Report on the Hazard of Near Earth Objects*, prepared for the UK government by a committee chaired by Dr. Harry Atkinson.

94 "... a Tunguska-type event wipes out Northern Italy ..." *Rendezvous with Rama*, by Arthur C. Clarke (1972).

94 "'Spaceguard'-type projects . . ." The relevant NASA report is at
   http://impact.arc.nasa.gov/reports/spaceguard/index.html.

94 ". . . to divert the trajectory . . ." As Carl Sagan noted, if it became
   feasible to change the orbits of asteroids, the technology could
   be used to divert them towards Earth rather than away from it,
   greatly increasing the natural "baseline" impact rate and turning
   asteroids into weapons, or instruments of global suicide.

95 "The Torino number assigned . . ." The Torino scale is described
   on http://impact.arc.nasa.gov/torino/.

96 ". . . a more refined index . . ." The Palermo scale was proposed in
   a paper by S.R. Chesley, P.W. Chodas, A. Milani, G.B. Valsecchi,
   and D.K. Yeomans, *Icarus* 159, 423–432 (2002).

Chapter 8

99 "The totality of life. . . " *The Future of Life*, by E.O. Wilson,
   Knopf, New York (2002).

101 "We are burning the books . . ." Robert May, *Current Science* 82,
    1325 (2002).

101 ". . . a Library of Life . . ." Gregory Benford's proposal is de-
    scribed in his book *Deep Time*.

102 ". . . 'footprint' needed to support each person . . ." The "foot-
    print" concept is discussed in the WWF "Living Planet Report"
    at http://www.panda.org.

105 "Almost ten percent . . ." These figures come from a recent
    report by NMG-Levy, a South African labour relations organisa-
    tion.

105 ". . . other calamitous 'natural' plagues . . ." Paul W. Erwald in
    The Next Fifty Years, Vintage Paperbacks (2002), John Brockman,
    ed., p. 289.

105 ". . . from decades to hundreds of millions of years." Five hun-
    dred million years ago, there was twenty times more carbon
    dioxide in the atmosphere than there is today: the greenhouse ef-
    fect was then far stronger. But the average temperature was not
    substantially higher in that era, because the Sun was intrinsically

fainter. The carbon dioxide started to fall when plants colonised the land, consuming this gas as the raw material for their photosynthetic growth. The gradual brightening of the Sun, a well-understood consequence of the way stars change as they get older, has counteracted the diminishing greenhouse effect, with the consequence that the mean global temperature has not changed much. There have however been fluctuations, between glacial and interglacial periods, of as much as ten degrees (centigrade) from the average value. Fifty million years ago, in the early Eocene geological era, there was still three times as much carbon dioxide in the atmosphere as there is today. There is fossil evidence for mangrove swamps and tropical forests in southern England at that time; the local temperature was then about fifteen degrees higher than it is now (though this was partly due to a shift in the continents and in Earth's spin axis, which placed England nearer the equator)

108 "... that trap the heat." This effect makes the Earth thirty five degrees hotter than it would otherwise have been. The key question is how many extra degrees of heating will be induced by human activities during this century.

109 "The anti-gloom environmental propagandist ..." The scientific issues regarding global warming are comprehensively discussed in the various reports of the Intergovernmental Panel on Climate Change (IPCC), on http://www.ipcc.ch.

111 "... 'conveyor-belt' flow pattern ..." A clear discussion of the "conveyor belt" concept is in W.S. Broecker, "What If the Conveyor Were to Shut Down? Reflections on a Possible Outcome of the Great Global Experiment," *GSA Today* 9(1):1–7 (January 1999). He notes that there have been sudden coolings in the past that if replicated, would transform Ireland's climate into that of Spitsbergen, turn Scandinavian forests into tundra, and freeze the Baltic Sea all the year round. He adds, however, that if there were a four- to five-degree warming before a human-induced "flip" occurred, the outcome, though still unpredictable, would be unlikely to be so extreme.

111 "... the next 'flip' much more imminent." *The Skeptical Environmentalist*, by Bjorn Lomberg, Cambridge University Press (2001).

111 "Earth would need to be substantially hotter than it actually is ..." Such a runaway could occur if the carbon dioxide level rose anywhere near to what it was 500 million years ago, the Sun being several percent brighter now than it was then. But the projected rise in carbon dioxide induced by human activities amounts to no more than a doubling—small compared to the twenty-fold changes that have occurred on geological timescales. In the natural course of events, the gradually-brightening Sun could trigger a runaway greenhouse effect due to evaporation from the oceans perhaps a billion years from now (even with present carbon dioxide levels). This could destroy land-based life far sooner than the more violent convulsions that accompany the Sun's death-throes 6 or 7 billion years hence. Greenhouse warming is even more drastic on the torrid planet Venus.

112 "... Charles, Prince of Wales ..." He was lecturing at Cambridge University in 1994, at the inauguration of a Global Security Programme at the University.

Chapter 9

116 "... the "precautionary principle." There is a huge literature on this subject. See, for instance, *Rethinking Risk and the Precautionary Principle*, edited by Julian Morris, Butterworth-Heinemann (2000).

117 "Edward Teller contemplated the scenario ..." *Memoirs: A Twentieth Century Journey in Science and Politics*, by Edward Teller, Perseus, p. 201 (2001).

117 "... a Los Alamos report." E. Konopinski, C. Marvin, and E. Teller, *Ignition of the Atmosphere with Nuclear Bombs*, Los Alamos Report. Until 2001 this was available on the Los Alamos website.

120 "... an experiment at Brookhaven ..." *COSM*, by Greg Benford, Avon Eos, NY (1998).

120 "... spatial dimensions beyond our usual three ..." See the comments on such theories in Chapter 11.

120 ". . . scientist produces a new form of ice . . ." *Cat's Cradle*, by Kurt Vonnegut, first published in 1963; available in e-version from Rosetta Books.

122 "Hut and I realised. . . ." Our paper was published as P. Hut and M.J. Rees, "How stable is our vacuum?" in *Nature* 302, 508–509 (1983).

123 ". . . asked a group of experts . . ." The Brookhaven report, entitled "Review of Speculative 'Disaster Scenarios' at RHIC," was published as R.L. Jaffe, W. Busza, J. Sandweiss, and F. Wilczek *Reviews of Modern Physics* 72, 1125-11–37 (2000).

124 ". . . summarised the situation like this:" The quotation is from S.L. Glashow and R. Wilson, *Nature* 402, 596 (1999).

125 ". . . a parallel effort . . ." The work of the CERN-based scientists A. Dar, A. de Rujula, and U. Heinz appeared as a paper entitled "Will Relativistic Heavy Ion Colliders Destroy our Planet?" in *Phys. Lett. B* 470, 142–148 (1999).

126 ". . . extinction cannot be felt . . ." Jonathan Schell in *The Fate of the Earth*, Knopf, New York (1982), pp. 171–172.

127 ". . . lack of candour in discussing. . . "Francesco Calogero's article "Might a Laboratory Experiment Now being Planned Destroy the Planet Earth?" is in *Interdisciplinary Science Reviews* 23, 191–202 (2000).

128 ". . . personal assessment of what was at stake." As I have emphasised in Chapter 3, we seem to have actually been exposed to a higher risk than most people realised—higher, I would guess, than any but the most fervent anti-Communists would have knowingly accepted.

131 ". . . what constitutes an acceptable risk . . ." Adrian Kent, "A critical look at catastrophe risk assessment," *Risk* (in press); preprint available as hep-ph/0009204.

Chapter 10

135 ". . . friend and colleague Brandon Carter." Carter's paper was published as "The anthropic principle and its implications for biological evolution," *Phil Trans R-Soc A* 310, 347.

136 "This Doomsday argument . . ." The most thorough critique of this line of argument is in *Anthropic Bias: Observation Selection Effects in Science and Philosophy*, by Nick Bostrom, Routledge, New York (2002). Another reference is C. Caves, *Contemporary Physics*, 41, 143–153 (2000).

138 "An even simpler argument was used . . ." J. Richard Gott III, Implications of the Copernican principle for our future prospects, *Nature* 363, 315 (1993) and his book *Time Travel in Einstein's Universe*, Houghton Mifflin, New York, (2001).

139 ". . . the Canadian philosopher John Leslie . . ." This argument is presented in Leslie's book *The End of the World: The Science and Ethics of Human Extinction*, Routledge, London (1996) (new edition 2000), which has a comprehensive account of hazards and the Doomsday argument. The author, a philosopher, brings zest to the gloomiest of themes. Further references to the Doomsday argument are given by Bostrom in his book cited earlier.

Chapter 11

141 ". . . John Horgan has claimed the latter . . ." Horgan's book *The End of Science* was published by Addison Wesley, NY, in 1996. An antidote is *What Remains to be Discovered*, by John Maddox, Free Press, New York and London (1999).

142 "No matter how much we learn . . ." The quotation, a response to a question by Heinz Pagels, is from *A Memoir*, by Isaac Asimov.

144 ". . . the quantum theory." Quantum theory wasn't the outcome of a single brilliant mind. Key precursor ideas were "in the air" in the 1920s, and the theory was pioneered by a remarkable cohort of young theorists, led by Erwin Schrödinger, Werner Heisenberg, and Paul Dirac.

144 "It is a tribute . . ." The quotation is from Stephen Hawking's *A Brief History of Time*, Bantam 1988.

144 "This theory is now confirmed . . ." As soon as the theory was proposed, Einstein realised that it explained some mysteries about the orbit of the planet Mercury. It was further confirmed

in 1919 by Arthur Eddington (one of my predecessors at Cambridge), who with colleagues measured how gravity deflected light rays passing near the Sun during a total eclipse.

145 "... limit to how precisely any clock can ever subdivide time." Even though there is as yet no theory of quantum gravity, the scales on which Einstein's theory must break down can be readily estimated. For example, the theory cannot consistently describe a black hole so small that its radius is less than the uncertainty in its position implied by Heisenberg's relation. This gives a minimum length of about 10^{-33} cms. The minimum quantum of time, known as the Planck time, would be this length divided by the speed of light, about 3×10^{-44} seconds.

145 "... a challenge for the twenty-first." This conceptual gap actually did not impede the huge twentieth-century advances in our understanding of the physical world, from atoms to galaxies. This is because most phenomena involve either quantum effects or gravity, but not both. Gravity is negligible in the micro world of atoms and molecules, where quantum effects are crucial. Conversely, quantum uncertainty can be ignored in the celestial realm, where gravity holds sway: planets, stars, and galaxies are so large that quantum "fuzziness" has no discernible effect on their smooth motions.

145 "currently the most favoured attempt at a unified theory ..." An accessible and entertaining summary of string theories and extra dimensions is *Strange Matters: Undiscovered Ideas at the Frontiers of Space and Time*, by Tom Siegfried, Joseph Henry Press (2002).

147 "Perhaps universes could be created. ..." This suggestion has been discussed by E.H. Fahri and A.H. Guth, (*Phys. Lett. B* 183, 149 (1987)) and by E.R. Harrison (*Q.J. Roy. Ast. Soc.* 36, 193 (1995)) amongst others.

149 "... little more chance than a fish." If physicists do discover a unified theory, it would be the culmination of an intellectual quest that started before Newton and continued through Einstein and his successors. It would exemplify what the great physicist Eugene Wigner called "the unreasonable effectiveness of

mathematics in the physical sciences." Also, if it is achieved by unaided human intellect, it would reveal that our mental powers can grasp the bedrock of physical reality, which would actually be a remarkable contingency.

152 "The big surprises . . ." The quotation is from John Maddox's book *What Remains to be Discovered* noted above.

152 ". . . midway between atoms and stars." In the prologue to this book I cited Frank Ramsey's personal perspective on the world: humans, the focus of his curiosity and concern, dominate the foreground; the stars are shrunk to relative insignificance. Science actually offers an objective rationale for this viewpoint, a viewpoint that is, of course, not peculiar to Ramsey, but is shared by almost all of us. Stars are (from a physicist's perspective) huge masses of glowing gas, squeezed and heated to immense temperatures by their own gravity. They are simple because no complex chemistry could survive the heat and pressure. A living organism, with layer upon layer of complicated internal chemistry, must therefore be far less massive than a star to avoid being crushed by gravity.

152 ". . . atoms in each of us." There are 1.3×10^{57} nucleons (protons and neutrons) in the Sun. The square root of this, 3.6×10^{-28}, corresponds to a mass of about fifty kilograms, within a factor of two of the mass of a typical human being.

155 ". . . using ever more powerful computers." The absolute theoretical limit to computer power, far beyond even what nanotechnology could achieve, has been discussed by the MIT theorist Seth Lloyd, who considers a computer so compact that it is on the threshold of becoming a black hole. See his paper "Ultimate physical limits to computation," *Nature* 406, 1047–1054 (2000).

Chapter 12

160 ". . . no techniques sensitive enough to disclose any of these faraway planets." The most successful current technique is an indirect one that involves detecting not the planet itself, but the

small wobble in the central star induced by the planet's gravitational pull. Jupiter-like planets induce motions of meters per second; Earth-like planets induce motions of merely centimetres per second, too small to be measured. But Earth-sized planets might reveal themselves in other ways. For example, if such a planet moved in front of a star, it would reduce its brightness by less than one part in ten thousand. The best hope of detecting this minuscule dimming would be to use a telescope in space, where the starlight is unaffected by Earth's atmosphere and therefore steadier. A planned European space mission called Eddington (named after the famous English astronomer) should be able to detect these transits of Earth-like planets across bright stars within the next decade.

160 ". . . Terrestrial Planet Finder." The tentatively favoured design—details are not yet finalised—would comprise four or five telescopes in space, arrayed as an interferometer in which the light from the star itself cancels out by interference (the peaks of the lightwaves reaching one telescope neutralising troughs from the lightwaves reaching the other) and so does not drown out the ultra-faint light from orbiting bodies.

161 ". . . possible 'twins' of our Earth." It is unclear what fraction of stars could have such a planet. Most of the planetary systems so far discovered are surprisingly different from our own solar system. Many contain Jupiter-like planets on eccentric orbits much closer in than our own Jupiter. These would destabilise any planet in a near-circular orbit at the "right" distance for its parent star to be an abode for life. We cannot yet be sure what fraction of planetary systems would permit a small Earth-like planet.

162 "Donald Brownlee and Peter Ward . . ." Their book *Rare Earth* is published by Copernicus, NY (2000).

164 "For all of life's plenitude . . ." The quotation is from Simon Conway Morris's article in *The Far Future Universe*, G. Ellis, ed., Templeton Foundation Press (Philadelphia and London 2002), p. 169. See also Conway Morris's book *The Crucible of Creation*, Cambridge University Press (1998).

165 "... a 'head start' on planets around these ancient stars." The astronomer Ben Zuckerman suggests (in *Mercury*, Sept–Oct 2002, pp. 15–21) another reason why we would expect visits if aliens existed. He points out that any aliens who had themselves surveyed the Galaxy with instruments like the Terrestrial Planet Finder would have identified the Earth as a specially interesting planet with an intricate biosphere long before humans came on the scene, and so had plenty of time to get here.

165 "... UFOs have indeed visited us." We should maybe be thankful to be left alone. An alien invasion might have the same effect on humanity as Europeans had on North American Indians and the islands of the South Pacific. *Independence Day* may be a truer depiction than *ET*.

167 "an entire language for interstellar communication, ..." Hans Freudenthal, *Lincos, a Language for Cosmic Intercourse*, Springer, Berlin (1960).

Chapter 13

169 "... this picture should be complemented by another one, ..." *The Fate of the Earth*, by Jonathan Schell p. 154.

174 "... advocated by the maverick engineer Robert Zubrin." The "Mars direct" strategy is described in *The Case for Mars: The Plan to Settle the Red Planet and Why We Must*, by Robert Zubrin with Richard Wagner, Touchstone (1996).

175 "Two years later ..." The relative positions of Earth and Mars are optimal once every two years. That is the reason why two years is the natural time interval between successive launches.

176 "... when propulsion systems are more efficient." This same problem would arise on any habitable planet, because gravity has to be this strong to retain an atmosphere at a temperature suitable for life.

177 "... what kind of novel propulsion systems will prove most promising." Solar panels can provide low thrust, for an unlimited time, in the inner parts of the solar system, but in the outer re-

gions sunlight is too weak, and even large and heavy panels yield very low power. At present, probes into deep space carry radioisotope thermoelectric generators (RTGs), which yield enough power for radio transmitters and other such equipment. To provide thrust for propulsion (especially if one requires enough to shorten journey times to the planets, rather than just midcourse corrections), some kind of nuclear fission reactor would be needed. This is a reasonable medium-term prospect. Longer-term and still speculative options include fusion reactors and even matter–antimatter reactors.

177 "Very thin carbon 'yarns' have already been made . . ." See K. Jiang, Q. Li, and S. Fan, *Nature* 419, 801 (2002).

179 ". . . Gerard O'Neill, an engineering professor at Princeton." O'Neill's ideas were published in the book *The High Frontier*, William Murrow, NY (1977), and promoted by an organisation called the "L5 Society." L5 denotes a position in the Earth–Moon system specially appropriate for locating a "habitat." G. Benford and G. Zebrowski's anthology *Skylife: Space Habitats in Story and Science* collects a set of fictional and scientific articles on this theme.

180 "Whether human beings will themselves have joined this diaspora . . ." This is one of Freeman Dyson's favourite themes, first adumbrated in his Bernal lecture. Indeed J.D Bernal in 1929 had ideas of this kind. A later Dyson reference is *Imagined Worlds*, Harvard/Jerusalem lectures (2001).

182 "The tree of life may reach twice . . ." *Life Evolving: Molecules, Mind and Meaning*, by Christian de Duve, Oxford University Press (2002).

182 ". . . look forward to a near-infinite future." In the 1960s, Arthur C. Clarke envisioned the "Long Twilight" after the death of the sun and today's other hot stars as an era at once majestic and slightly wistful. "It will be a history illuminated only by the reds and infrareds of dully glowing stars that would be almost invisible to our eyes; yet the sombre hues of that all-but-eternal universe may be full of colour and beauty to whatever strange beings

have adapted to it. They will know that before them lie, not the ". . . billions of years that span the past lives of the stars, but years to be counted literally in trillions. They will have time enough, in those endless aeons, to attempt all things and to gather all knowledge. But for all that, they may envy us, basking in the bright afterglow of creation; for we knew the universe when it was young." (reprinted in *Profiles of the Future*, Warner Books, NY (1985))

Chapter 14

185 ". . . dated the creation at Saturday afternoon . . ." An accessible summary of Archbishop Ussher's life and work, and of progress towards our modern chronology, is in *Aeons*, by Martin Gorst, Fourth Estate, London (2001). Ussher's chronology, starting with the creation in 4004 B.C.E., was featured until 1910 in the Bibles published by the Oxford University Press.

188 ". . . in this great Celestial Creation, . . ." From Thomas Wright of Durham's *An Original Theory or New Hypothesis of the Universe* (1750), reprinted by Cambridge University Press with an introduction by Michael Hoskin. Wright goes on to set worldly troubles in a cosmic perspective more relaxed than most of us could share: "I can never look upon the Stars without wondering why the whole World does not become Astronomers . . . and reconcile them to all those little Difficulties incident to human Nature, without the least Anxiety."

INDEX